MEANS TO AN END
Apoptosis and Other Cell Death Mechanisms

MEANS TO AN END
Apoptosis and Other Cell Death Mechanisms

DOUGLAS R. GREEN
St. Jude Children's Research Hospital

COLD SPRING HARBOR LABORATORY PRESS
Cold Spring Harbor, New York • www.cshlpress.com

MEANS TO AN END
Apoptosis and Other Cell Death Mechanisms

© 2011 by Cold Spring Harbor Laboratory Press, Cold Spring Harbor, New York
All rights reserved
Printed in United States of America

Publisher	John Inglis
Acquisition Editor	Richard Sever
Project Manager	Maryliz Dickerson
Production Editor	Rena Steuer
Desktop Editor	Susan Schaefer
Book and Cover Designer	Denise Weiss

Front cover: ©2010 Artists Rights Society (ARS), New York/ADAGP, Paris. Max Ernst (1891–1976). The gramineous bicycle garnished with bells, the dappled fire damps, and the echinoderms bending the spine to look for caresses. 1920 or 1921. Botanical chart altered with gouache, 29 ¼ x 39 ¼ inches. Purchase. (279.1937.) The Museum of Modern Art, New York, N.Y. Photo Credit: Digital Image ©The Museum of Modern Art/Licensed by SCALA/Art Resource, New York.

Library of Congress Cataloging-in-Publication Data

Green, Douglas R.
  Means to an end : apoptosis and other cell death mechanisms / Douglas R. Green.
      p. cm.
   Includes bibliographical references and index.
   ISBN 978-0-87969-887-4 (hardcover : alk. paper) -- ISBN 978-0-87969-888-1 (pbk : alk. paper)
  1. Cell death. 2. Apoptosis. I. Title.

   QH671.G74 2011
   571.9'36--dc22

                                     2010033398

10 9 8 7 6 5 4 3 2 1

Authorization to photocopy items for internal or personal use, or the internal or personal use of specific clients, is granted by Cold Spring Harbor Laboratory Press, provided that the appropriate fee is paid directly to the Copyright Clearance Center (CCC). Write or call CCC at 222 Rosewood Drive, Danvers, MA 01923 (508-750-8400) for information about fees and regulations. Prior to photocopying items for educational classroom use, contact CCC at the above address. Additional information on CCC can be obtained at CCC Online at http://www.copyright.com/

All Cold Spring Harbor Laboratory Press publications may be ordered directly from Cold Spring Harbor Laboratory Press, 500 Sunnyside Boulevard, Woodbury, New York 11797-2924. Phone: 1-800-843-4388 in Continental U.S. and Canada. All other locations: (516) 422-4100. FAX: (516) 422-4097. E-mail: cshpress@cshl.edu. For a complete catalog of Cold Spring Harbor Laboratory Press publications, visit our World Wide Web Site http://www.cshlpress.com/

*To my parents, Lila and Warren Green,
who gave me their passion for exploring nature*

Contents

Foreword by Martin Raff, ix

Preface, xi

Introduction, 1

1   A Matter of Life and Death, 5
2   Caspases and Their Substrates, 15
3   Caspase Activation and Inhibition, 29
4   The Mitochondrial Pathway of Apoptosis, Part I: MOMP and Beyond, 47
5   The Mitochondrial Pathway of Apoptosis, Part II: The BCL-2 Protein Family, 63
6   The Death Receptor Pathway of Apoptosis, 87
7   Other Caspase Activation Platforms, 99
8   Nonapoptotic Cell Death Pathways, 111
9   The Burial: Clearance and Consequences, 127
10 Cell Death in Development, 149
11 Cell Death and Cancer, 163
12 The Future of Death, 177

Figure Credits, 191

Additional Reading, 193

Index, 215

Foreword

Biologists have been relatively slow to recognize and study the degradation processes that operate in cells, compared with the generative processes. Our understanding of protein degradation, for example, lagged well behind our comprehension of protein synthesis. And so has it been for cell death, where understanding followed many years behind the comprehension of cell division. Although it had long been recognized that cell death can be an important part of normal animal development and tissue homeostasis, it was only in 1972 that Kerr, Wyllie, and Currie drew a clear distinction between the conserved cytological features of these normal cell deaths and the very different features of acute pathological cell deaths. They coined the term apoptosis for the former type of cell death and, importantly, suggested that it might reflect the operation of a conserved intracellular death program, by which animal cells can actively kill themselves in a tidy and controlled way.

This important idea remained largely dormant for almost 20 years, and the study of apoptosis remained confined to a small group of aficionados working on diverse organisms. The big bang in the cell death field came from Horvitz and colleagues at the end of the 1980s and early 1990s with the genetic identification of the intracellular proteins that mediate and regulate apoptosis in the nematode *Caenorhabditis elegans* and, soon thereafter, the demonstration that related proteins operate in similar ways in other animals, including humans. In this way, it rapidly emerged that a family of cysteine proteases—the caspases—mediate the apoptotic death program and that a family of regulatory proteins—the BCL-2 proteins—either activate or repress the program. These spectacular findings indicated that apoptosis is a fundamental property of animal cells and that the proteins that mediate and regulate it have been largely conserved in evolution from worms to humans. The findings launched the subject into the cell biological stratosphere, where it remains to this day, having gone from neglect to hysteria in only a few years.

As the cell death field matured, it became increasingly clear that there are multiple ways of activating and repressing the apoptotic program from both inside and outside the cell. It also emerged that the molecular details can vary from organism to organism and that other nonapoptotic death programs can operate in animal cells. This added complexity has created a pressing need for a comprehensive stock taking—a cool, clear, overview of cell death that cuts through the detail in a logical and engaging way while making it clear where controversy and mystery remain. The author of *Means to an End*, a highly respected leader in the field, has achieved all of this admirably. The writing is remarkably clear and is bolstered by simple, informative figures.

Whether you are a cell death expert or a neophyte, or even a retired cell biologist like me, you are likely to find the book informative, clarifying, and enjoyable. If you are a scientist just starting your career in the cell death field, it is unlikely that you will find a better place to identify important unsolved problems on which to work. If you are a drug developer, you will find an enlightened discussion of how one might design drugs to either encourage dangerous cells to kill themselves or discourage transiently injured cells from doing so. All you need to know about cell death is covered here, with panache, and all in fewer than 250 pages—a remarkable achievement.

MARTIN RAFF
London, July 2010

Preface

This book is not a text book, nor is it a monograph. It is not a history of the field of cell death, nor is it an exhaustive treatment. What it is *intended* to be is a starting point for those who are interested in cell death, especially in mammals, a subject about which I am admittedly passionate. The field has reached an interesting point at which I believe we have a reasonable understanding of the processes that are central to or consequences of cell death and its regulation. The book might be considered an argument that we have, indeed, reached this point.

I wrote this book primarily for students and informed individuals who would like to learn more about cell death. A college level understanding of cell biology and biochemistry is assumed, but I am not sure that it is absolutely essential. I have tried to define terms and processes that are necessary to comprehend the material. But there is certainly a limit: You must have a fairly clear idea of what a protein is, the parts of a cell, and cell signaling and transcription. That said, I have also tried to write the book so that it is of interest to those who work in the field (who will undoubtedly discover my mistakes, biases, and eccentricities).

This book is meant to be read from beginning to end. The order of the chapters is intentional, so that concepts developed in one chapter are built upon in chapters that follow. It is unlikely that any reader other than an expert will learn enough from any given chapter to gain anything close to a complete understanding, but if it is put into the context of the chapters that precede it, there should be many "aha" moments. The illustrations should be helpful, but they do not contain all the information that you need—that is in the text.

At the end of the book is a list called "Additional Reading." This list serves two purposes. First, it provides sources of further information and alternative viewpoints. Second, many (but not all) of the landmark papers upon which I based the positions I have taken are included.

As the author, I am responsible for the content of the book, including the mistakes and misconceptions. But under no circumstances should it be thought that I had anything to do with the vast majority of the discoveries on which the concepts, ideas, notions, and musings are based. I admit that I do draw attention to some discoveries that were made in my own lab, by people I am proud of, and I have tried not to do this unduly.

The field of cell death owes a tremendous debt to many pioneers, and I have not given them suitable credit in the text for their phenomenal contributions. In particular, the field as we know it was the work of pathologists who recognized that there are discrete forms of cell death, scientists working in invertebrate models (especially nematodes and flies) who unraveled the genetic basis of cell death, and cancer biologists who identified molecules that control cell death. As I mention above, this book is not a history of the field. However, some of those pioneers who directly influenced my views and understanding of the field bear mention and thanks (in no particular order): Andrew Wyllie, John Kerr, John Cohen, Sten Orrenius, Stanley Korsmeyer, Yoshihide Tsujimoto, Jerry Adams, Martin Raff, Suzanne Cory, David Vaux, John Reed, Junying Yuan, Guido Kroemer, Andreas Strasser, Shigekazu Nagata, Peter Krammer, Michael Hengartner, Xiaodong Wang, David Wallach, Gerard Evan, Jurg Tschopp, Yigong Shi, Guy Salvesen, Tak Mak, Herman Steller, Hao Wu, Michael Karin, Richard Youle, Emad Alnemri, Vishva Dixit, Gerry Melino, and all of the others that I may have inadvertently missed (my apologies).

I must also mention some of those who made this book possible: Richard Sever, my editor and tormentor (just kidding), who cajoled me into writing this, and then stood steadfastly by me throughout; Tudor Moldoveanu, who generated many of the figures showing molecular structures; Guy Salvesen, Eric Baehrecke, and Carlos Lopez who read selected chapters and provided advice; Lisa Bouchier-Hayes and Melissa Parsons who read the entire book, corrected my errors, questioned some of my arguments, and helped enormously; and my family, Rona Mogil and Maggie Green, who put up with this project and with me.

I also would like to thank my parents, who inspired my love of science, and to whom this book is dedicated. When my sister and I were growing up, we spent our family time in the forest, learning to identify trees, birds, and animal tracks, and we thought that it was perfectly normal to veer off highways to collect wild flowers to plant in the yard. Although our parents are not scientists, we both became biologists, and my sister studies penguins on her base in the Antarctic. I chose to study cells, in part because I like to live where one can get a good martini.

I hope you enjoy reading this. Those of us who do science for a living know that often science is very hard work, but it can also be enormous fun. Most of all, I hope that the book encourages you to experience the latter.

DOUG GREEN
Memphis, July 2010

Introduction

Like all living things, cells die. Indeed, a great many cells in our bodies die throughout our lives, and their deaths are essential for our survival. They die by highly conserved mechanisms that may have their evolutionary origins more than 1 billion years ago. This book is about how that death happens and how it contributes to physiological homeostasis and disease. The focus is on cell death in animals and predominantly on only one form of cell death, called *apoptosis,* for two reasons. First, most cells that die in humans die by apoptosis. Second, it is the type of cell death about which we currently know most. So, although other types of cell death are covered in some detail, most of our discussion concerns apoptosis.

The underlying mechanisms of apoptotic cell death are found throughout the animal kingdom, but probably nowhere else.[1] In all animals studied (including many of the phyla), the features of this type of cell death are the same: The dying cell effectively "packages" itself to be eaten by healthy cells and digested. Furthermore, many of the specific molecules involved in this process are conserved in animals. However, the specific molecular pathways, although similar, can have fundamental differences. Throughout most of this book, we focus on the molecular pathways of apoptosis (and cell death in general) that function in humans. This unabashedly anthrocentric (or, at least, "backbonecentric") view is our goal, with apologies in advance to those readers who consider themselves "fly people" or "worm people" and those with interests in other organisms.[2]

---

[1] There is a literature that explores cell death in other types of organisms, including plants and yeast, and it remains possible that within the cell death mechanisms in such organisms is a vestige of a far more ancient process than we currently suspect. However, the molecules involved in the process are, at best, very distantly related to those in animals, and the actual pathways remain to be elucidated.

[2] In fact, it is largely owing to those who study such organisms (especially nematodes and insects) that we know so much about the molecular mechanisms of apoptosis, and hence the apology (however flippant it may appear) is meant with sincerity. Robert Horvitz was awarded a Nobel Prize in Medicine and Physiology in 2002 for his pioneering studies of cell death in nematodes.

We will quickly step off into the deep end of the molecular pool with the biochemical mechanisms of cell death. For reasons that will become clear, a "bottom-up" view of cell death by apoptosis comprises the first several chapters. But before we dive in, it may be useful to say a word about how the chapters that follow are organized.

- Chapter 1 is essentially a synopsis, a quick take on the rest of the book. It is a chance to get our bearings and take a stab at the big picture, starting with why cells die and the three major types of cell death. It goes on to outline the molecular mechanisms covered in subsequent chapters.
- Chapters 2 and 3 concern caspases, the proteases that orchestrate apoptosis by cleaving substrates in the cell. Chapter 2 introduces the caspases and explores those substrates that have known roles in apoptosis. We also discuss caspases that are not involved in apoptosis, per se. Chapter 3 considers the biochemistry of activation of different types of caspases, as well as inhibitors of caspases and their roles.
- Chapters 4 and 5 cover the mitochondrial pathway of apoptosis, the major way in which apoptosis occurs, at least in vertebrates. Chapter 4 discusses the events that occur once the mitochondrial pathway is engaged and how this leads to caspase activation. It also introduces the caspase activation pathways of flies and nematode worms, together with ideas on how apoptosis may have evolved. Chapter 5 introduces the BCL-2 family of proteins, whose complex interactions link different signals for cell death to the mitochondria.
- Chapter 6 considers another way apoptosis is engaged in vertebrates—by cell surface death receptors—and how these specialized receptors engage a distinct pathway of caspase activation. This pathway can also link to the mitochondrial pathway to cause apoptosis.
- Chapter 7 looks at additional pathways for caspase activation. One is engaged by signals from infectious organisms and some inert substances and can result in a form of apoptosis but also triggers inflammatory responses. Another pathway involves the most highly conserved of the caspases and how it is activated, but its role in cell death is obscure.
- Chapter 8 explores the other major forms of cell death, necrosis, and autophagic cell death. Chapter 9 covers what happens after a cell dies. Regardless of how it died, a dead cell is rapidly cleared from the body by phagocytosis. But once the cell is gone, there are additional consequences, including effects on the immune system and proliferation of healthy cells.
- Chapter 10 provides examples of cell death in development, exploring how the cell death pathways can be engaged by signals that specify which cells must die. Chapter 11 introduces the idea that cancer is, in part, a disease of defective cell death. It discusses the mechanisms that are in place to prevent cancer and how

these link to the machinery of apoptosis, as well as the roles that cell death may have in promoting cancer and in cancer therapy.

- Chapter 12 explores the mechanisms of cell death as we understand them and how these are tested. These include formal models and their consequences for biology, as well as the practical applications of these mechanisms to the treatment of disease.

Chapter 1

A Matter of Life and Death

CELL DEATH IS ESSENTIAL

Every second, something on the order of 1 million cells die in our bodies.[1] This is a good thing, because cell death is central to proper homeostasis and adaptation to a changing environment. When, for some reason, it does not occur, the consequences can be catastrophic, manifesting as cancer, autoimmunity, or other maladies. Alternatively, if cell death occurs at the wrong time and place, this can also produce untoward effects, such as stroke, degeneration, heart attack, and many other injuries. Cell death is a major component of disease.

Cells die for a variety of reasons. For example, they can become physically or chemically stressed to the point that they cannot maintain their integrity. More often, however, such stresses engage an active cell death process, or apoptosis, before the stress becomes overwhelming; that is, the cell essentially commits suicide. Most, but not all, cell suicide occurs by apoptosis.

Another stress that triggers cell death is infection of the cell by an organism that would use the cell for its own parasitic ends (i.e., to make additional parasites). Such cell death can protect the body from further infection, unless the parasite can prevent this protective death until it has reproduced. In organisms with immune systems, lymphocytes and phagocytic cells can detect an infected cell and instruct it to die.

But it is not only in response to stress that cells die. In the course of normal development, cells may serve functions that become superfluous, and developmental cues (in the form of specific molecules) can signal a cell to die. Technically, the death of a cell at a prescribed time in development is called *programmed cell death*, although this

---

[1] We do not know the exact number, but if we consider only one type of cell, the neutrophil, it is estimated that up to 1 million of these cells die per second. The turnover of mucosal epithelium (in the lung and gut) is also very high. One million per second is probably an underestimate, but it is in the ballpark.

term is often applied to apoptosis in general, probably incorrectly.[2] Another situation in which cell death occurs is when a cell acquires a mutation that allows it to lose its social inhibitions and proliferate as a cancer. Tumor-suppressor mechanisms that are built into our cells instruct them to undergo suicide rather than imperil the body at large.

TYPES OF CELL DEATH

There are three major types of cell death[3]: apoptosis, autophagic cell death, and necrosis. *Apoptosis* (sometimes called type I cell death) means, roughly, a "falling off."[4] The term was coined to invoke leaves falling from a tree, because usually, cells that die by apoptosis do so in tissues in an apparently random manner.

When cells die by apoptosis, the contents of the dying cell remain contained in membranes. The plasma membrane contorts into "blebs," and often the cell fragments into smaller membrane-bound *apoptotic bodies* (this term is also applied to apoptotic cell corpses that do not break up). The nucleus undergoes characteristic changes, including chromatin condensation, and it, too, often breaks up. Another characteristic feature of apoptosis is the cleavage of DNA, usually into pieces that are multiples of nucleosomes that appear as a "ladder" on agarose gels. Other organelles do not undergo dramatic morphological changes. The dying cell shrinks, and if it is adherent, it detaches from surrounding cells. Some of the morphological changes associated with apoptosis are shown in Figure 1.1.

Apoptotic bodies are rapidly removed by other healthy cells through phagocytosis. In a relatively short time, nothing obvious remains to indicate that a cell had ever been there. Generally, apoptosis is not associated with a subsequent inflammatory response (but it can be).

Autophagic cell death (type II cell death) differs from apoptosis and is less well understood. It involves a cellular mechanism of self-eating, or *autophagy*. Whether autophagy kills the cell is controversial, and most, but not all, studies suggest that autophagy accompanies such cell death in a last-ditch effort to keep the cell alive. Often, this form of cell death occurs when, for some reason, apoptosis is blocked. Autophagic cell death is characterized by the appearance of vacuoles in the cell and does not include extensive condensation of the nucleus (Figure 1.2).

---

[2] A distinction between *programmed cell death* and *apoptosis* is worth making, because not all apoptosis is "programmed" by genetic events, and not all developmentally programmed cell death occurs by apoptosis. But this is a losing proposition, because a great many people use the terms interchangeably. In this book, we avoid the use of "programmed cell death" unless we are discussing specific developmental events.

[3] This is based on recommendations made by the Nomenclature Committee on Cell Death (in 2009).

[4] Although it is often related that the first use of the term comes from Homer, a careful perusal of that esteemed author's works by Mauro Degli-Esposti failed to turn up this word. Instead, the first use of the term appears to be in the works of Hippocrates, who used it in a medical context 2400 years ago, referring to the "falling off of bones" in gangrene. It was later used by Galen as well and extended to the "falling off of scabs."

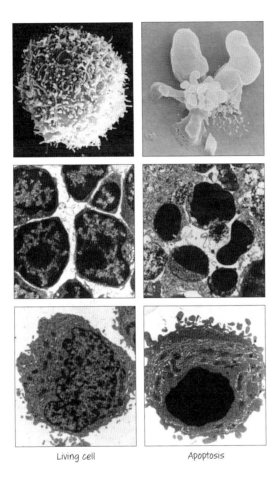

Living cell   Apoptosis

Figure 1.1. Apoptosis snapshots.

The third major type of cell death, *necrosis* (also known as type III cell death), is a messy sort of death in which the cell swells, bursts, and decomposes. It can occur when the plasma membrane is ruptured or when energy levels drop so quickly that the cells cannot sustain themselves (Fig. 1.3). However, there are forms of necrosis that are "programmed," that is, cells contain suicide pathways that result in necrosis rather than apoptosis.

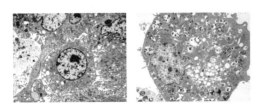

Figure 1.2. Autophagic cell death.

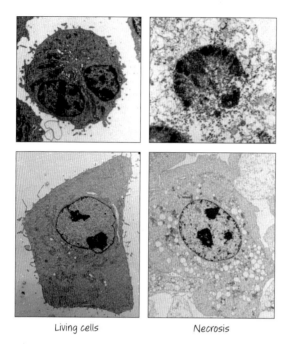

Figure 1.3. Necrosis.

There is another form of cell death worth mentioning, but it is very specialized. When skin is created, keratinocytes die by a process called *cornification*: The nucleus is degraded and the proteins of the cell are extensively cross-linked to make a hard, dead material that has a very important barrier role in our bodies. This form of cell death is distinct from those we consider in this book, and because it is a very specialized process, we do not discuss it further.

There are many other types of cell death that have thus been described with terms such as pyroptosis, necroptosis, aponecrosis, paraptosis, mitotic catastrophe, etc. In general, these are not sufficiently well characterized to qualify them as unique forms of cell death and we generally include them under the three major types of cell death outlined above.

APOPTOSIS FROM THE "BOTTOM UP"

Apoptosis is most easily understood by viewing it in reverse, beginning at the extreme end of a cell's life and tracking back to the healthy cell and what caused it to die. The "packaging" of the dying cell and the signals it sends to other cells to clear it away are brought about by the proteolytic cleavage of many hundreds of different proteins in the cell. The proteases responsible for this cleavage are called *caspases* and, in particular, they are referred to as a subset called *executioner caspases*.

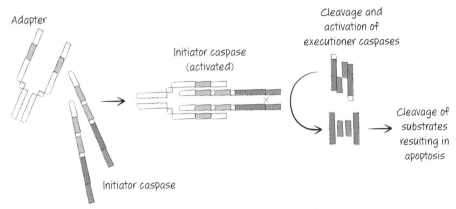

Figure 1.4. General caspase activation pathways.

Executioner caspases are already present in healthy cells but in an inactive form. They become activated when they are themselves cut by other proteases, called *initiator caspases*. These, too, are present in healthy cells, but their activation mechanism is different. Initiator caspases are inactive until two identical chains are brought together by *adapter* proteins to produce an active enzyme ("adapter" is a general term used to describe proteins that bind to and bring together other proteins; we use the term here to refer to those that perform this function for the initiator caspase chains).

Different adapter proteins are engaged by different apoptotic stimuli and define distinct *apoptotic pathways* (Fig. 1.4).

When adapters promote the activation of initiator caspases, the latter activate the executioner caspases. The executioner caspases then cleave hundreds of substrates and the cell undergoes apoptosis. The different pathways converge on the same set of executioner caspases, which is why our bottom-up view is warranted and is the way in which we unravel these pathways in subsequent chapters of this volume.

ROADS TO RUIN: THE PATHWAYS OF APOPTOSIS

Most apoptosis in vertebrates (at least) occurs by what is termed the *mitochondrial pathway*. In this pathway, conditions that promote cell death engage a set of related proteins, called the BCL-2 family (Fig. 1.5), that control the integrity of the outer membranes of mitochondria in the cell. Pro-apoptotic BCL-2 effectors disrupt outer mitochondrial membranes, whereas anti-apoptotic BCL-2 proteins prevent this and thereby prevent apoptosis. A third set of BCL-2 proteins regulates the other two types.

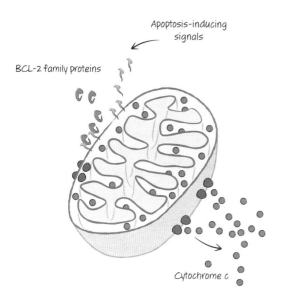

Figure 1.5. Simplified scheme of the first part of the mitochondrial pathway. BCL-2 proteins control mitochondrial outer membrane permeabilization (MOMP), releasing proteins that include cytochrome c.

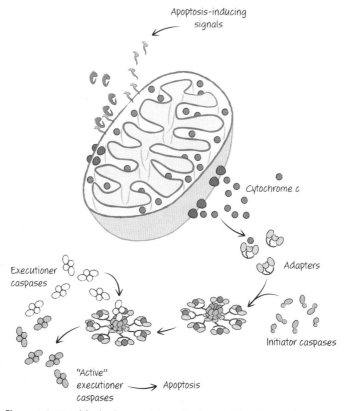

Figure 1.6. Simplified scheme of the mitochrondrial pathway of apoptosis.

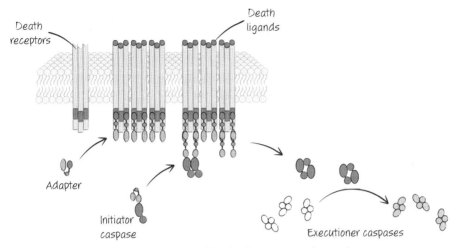

Figure 1.7. Simplified scheme of the death receptor pathway of apoptosis.

If mitochondrial outer membrane permeabilization (MOMP) occurs, soluble proteins of the intermembrane space (between the outer and inner mitochondrial membranes) diffuse into the cytosol. These proteins include cytochrome c, which also has a central role in mitochondrial physiology.

When cytochrome c reaches the cytosol, it interacts with an adapter protein that is present there, causing it to cluster (oligomerize) and bind to one of the initiator caspases. This then becomes active (Fig. 1.6) and, in turn, cleaves and thereby activates executioner caspases that cleave their substrates. Apoptosis ensues.

A second pathway for caspase activation and apoptosis involves specialized receptors on the cell surface called death receptors. When the ligands for these receptors bind, the intracellular region of these receptors engage a specific adapter molecule (distinct from that of the mitochondrial pathway). This, in turn, binds to and thereby activates an initiator caspase (again, different from that of the mitochondrial pathway). The initiator caspase cleaves and thereby activates executioner caspases (the same as above) and apoptosis proceeds (Fig. 1.7).

As mentioned above, another type of signal that can induce apoptosis is infection of a cell. Our cells have sensors that detect such infection, and some of these also act to engage adapters (distinct from those of other pathways) for a type of caspase related to initiator caspases. In addition to activating executioner caspases and causing apoptosis, this caspase also processes and allows the secretion of mediators that engage host defenses to fight the infection (Fig. 1.8).

The details of these apoptosis pathways are discussed in much more detail in the rest of this book, but the basics of the pathways are as described here. As we will see, apoptosis in other animals also follows these general schemes of adapter-caspase interactions.

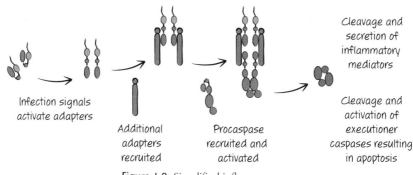

Figure 1.8. Simplified inflammasome.

THE "TOP" OF THE PATHWAY: WHAT INDUCES APOPTOSIS?

The pathways we have briefly considered are triggered by distinct types of signals. Most events that stress a cell engage the mitochondrial pathway of apoptosis. These include loss of growth factors, agents that damage DNA, disruption of the cytoskeleton, extensive protein aggregation, and other forms of cell stress. This pathway is also often triggered by developmental signals and by tumor-suppression mechanisms. The death

---

A GENERAL PROBLEM IN CELL DEATH RESEARCH

What death is may seem obvious. But in practice, cell death turns out to be a bit tricky to define in a general way. If a cell loses its plasma membrane integrity, it is certainly dead, but as we mentioned above, apoptotic cells are often engulfed before this event, and a cell that has been eaten and is being digested is dead. However, if we wait until that event to register this as a "cell death," we face a kinetic problem, because some cells will be long gone before others engage the pathway, and any count of dying cells may be an underestimate. Therefore, we might prefer to call a cell "dead" if it has activated caspases and is undergoing apoptosis. However, here we have a problem, because it often happens that cells that initiate apoptosis die even when caspases are not activated, although the death does not have the features of apoptosis. Alternatively, we might count a cell as "dead" if it loses the ability to proliferate or perform its functions. However, cells that lose the ability to proliferate can persist for years (indeed, this is very important for many of our cells), and functions change. So, when is a cell "dead?"

In practice, the definition of "death" is often tied to the cells and systems under study. Therefore, surrogate markers of cell death are often used, and it is assumed that these represent cellular demise. Although extensive progress has been made using this approach, mistakes abound. Undoubtedly, possible mistakes in this book may be the result of the interpretation of such surrogates as "death," and it is useful to keep this in mind.

receptor pathway is engaged by the specific ligands that bind to the receptors. The inflammasome pathway is induced by signals resulting from infection, although some inert substances can act in the same way if taken up into the cell. Therefore, apoptosis is engaged by a wide range of conditions, and there can be extensive overlap in which pathway can be triggered by a specific agent, depending on the type of cell. As we will also see, these pathways can also "cross talk," in that the death receptor pathway or inflammasome pathway can engage the mitochondrial pathway as well.

DYING CELLS ARE SIGNALS

When a cell dies, other cells in the body rapidly remove the corpse.[5] Cells that undergo apoptosis produce signals, as a consequence of caspase activation, that attract cells capable of eating them and other signals that cause their removal before the plasma membrane is disrupted. Necrotic cells, in contrast, release intracellular molecules that also lead to their clearance, but in a way that promotes inflammation of the tissue. Depending on how the cell died, our immune system can therefore be engaged to respond to whatever might have caused the cell death. Alternatively, it can be instructed *not* to respond.

The way in which dying cells affect the body are complex, and the consequences of cell death are varied. When a cell is gone, it may not be forgotten.

---

[5] We seem to be violating our "bottom-up" view by discussing the consequences of cell death (admittedly, the "bottom") only after discussing the "top" of the pathway. But cells that die by any mechanism, not only apoptosis, can have an impact on physiology.

CHAPTER 2

Caspases and Their Substrates

KILLER PROTEASES

Apoptosis is orchestrated by a set of proteases called caspases that reside in inactive form in nearly all of our cells. When activated, some caspases cleave hundreds, or perhaps thousands, of specific proteins. Such cleavage leads to all of the features that characterize this type of cell death, and apoptosis depends on the functions of caspases. In this chapter, we introduce these enzymes and the substrates that they cut to bring about apoptosis.

Caspases are endopeptidases; that is, they cut proteins internally rather than nibble away at the proteins' ends. Unlike digestive proteases such as trypsin, caspases do not degrade their substrates, but rather, they clip them at discrete sites. In general, they cut specific sequences that end in aspartate residues and cut immediately after this amino acid (this is the "asp" in caspase). However, caspases do not cut after every aspartic acid that can be accessed in a protein; they have more sequence specificity.

In caspases, the active site of the enzyme includes a cysteine, which therefore classifies caspases as *cysteine proteases* (this is the source of the "c" in caspase: a cysteine protease that cuts proteins after aspartic acid residues).

HOW CASPASES CUT PROTEINS

Cysteine proteases, also called thiol proteases, are one of the four major types of proteases characterized by their active sites (the others are serine, aspartyl, and zinc proteases). Although the focus here is on how caspases work, there are similarities among all proteases.

For a caspase to work, the interaction with its protein substrate must be "fast-on–fast-off." This involves the enzyme briefly holding the target peptide bond at the active site.

In a functionally active caspase, a substrate specificity pocket is close to the active cysteine, which is itself near a histidine. This cysteine-histidine diad is where the action takes place. The critical residue in the substrate pocket is an arginine, which holds

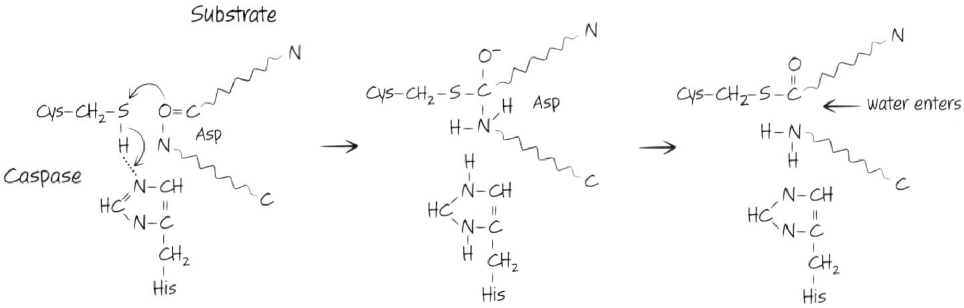

Figure 2.1. The acylation step of caspase cleavage.

the target aspartate in the substrate in position. This results in the situation shown in Figure 2.1, and the reaction follows. This step is called acylation; the next step is deacylation, shown in Figure 2.2. A water molecule is sacrificed, the peptide bond after aspartate is cut, and the caspase is now ready to cut another substrate protein.

For a caspase to function, the catalytic diad, cysteine histidine, must be brought close to a target protein. This is the function of the substrate specificity pocket in the caspase. That pocket, with its arginine, only permits aspartates (and, to a much weaker extent, glutamate, in some cases) to gain proximity to the diad. Inactive caspases (procaspases), however, do not allow access to the catalytic diad. To form the pocket, changes to the procaspase must occur to form an active site. How this occurs depends on the type of caspase.

TYPES OF CASPASES

Caspases, like apoptosis, are found only in animals,[1] and in most animals, there are several different caspases. In general, types of caspases can be distinguished by (1) their functions, (2) structure of the proforms, and (3) how they are activated. These are often interrelated, and the distinctions are not absolute.

In vertebrates and probably most other animals, at least two types of caspases are involved in apoptosis (the notable exception is nematodes, as discussed below): *executioner* caspases (such as in mammals, caspase-3, -6, and -7) and *initiator* caspases (in mammals, caspase-8 and -9). Initiator caspases are sometimes referred to as "apical" caspases. Another type of caspase, related to initiator caspases, includes *inflam-*

---

[1] Proteases are found throughout living things. Although caspases are only found in animals, proteases can have structural (but not sequence) similarity to caspases. Some of these have been called metacaspases and paracaspases. Some studies have suggested that these proteases can have roles in cell death in plants and fungi, but at present, the details of these processes remain obscure. Metacaspases are found in animals but are not known to have roles in apoptosis.

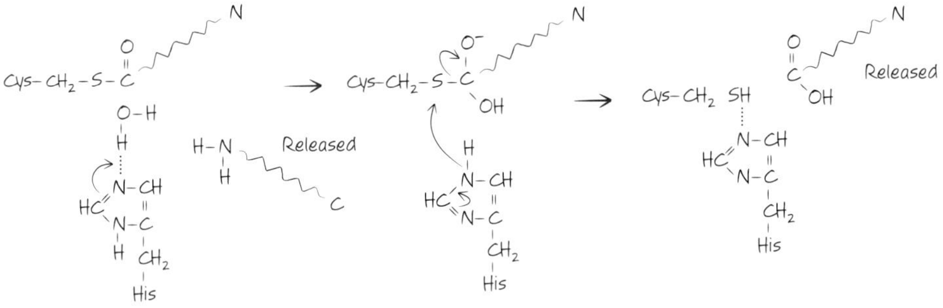

CASPASES AND THEIR SUBSTRATES   17

Figure 2.2. The deacylation step of caspase cleavage. The mechanism is based on the function of other cysteine proteases and has not been confirmed in caspases.

matory caspases (in humans, caspase-1, -4, and -5; in rodents, caspase-1 and -11). Additional caspases (in mammals, caspase-2, -10, -12, and -14, among others) are more difficult to place into one of these categories; their functions are less understood.

All procaspases have three regions, termed the prodomain, large subunit, and small subunit. The latter two are always separated by one or more sites, where they are cleaved by the action of the caspase itself, as it becomes activated, or by another protease. Schematic examples of several human caspases are shown in Figure 2.3.

The prodomain and large subunit are often separated by a cleavage site targeted by the caspase itself. The functions of these cleavage sites vary in different caspases,

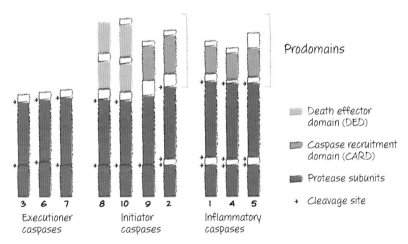

Figure 2.3. Schematics of several human caspases. Caspase-2 and -10 are grouped with the initiator caspases, although their classification is problematic.

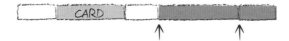

Figure 2.4. CED-3 linear organization.

as discussed in Chapter 3. For now, it is sufficient to be aware that there are different types of caspases with different biochemistries and functions. Those that concern us in this chapter are primarily executioner caspases, which, in vertebrates, are caspase-3, -6, and -7. These caspases become active during apoptosis and cleave substrates that bring about the characteristic features of apoptosis.

CASPASES IN OTHER ANIMALS

Nearly all animals for which we have genomic information appear to have caspases, and in some invertebrate animals, we know that executioner caspases are important for apoptosis. As mentioned in the Introduction, the pathways of apoptosis in two invertebrates, namely, the nematode *Caenorhabditis elegans* and the fruitfly *Drosophila melanogaster*, have been characterized in detail.

In *C. elegans*, only one caspase, CED3, is important in apoptosis, and CED3 has features of an initiator caspase (e.g., a long prodomain) but also acts as an executioner caspase, orchestrating apoptosis in the dying cells (Fig. 2.4).

In flies, several different types of caspases have been identified, and these are shown in Figure 2.5. Of these, DCP1 and Drice appear to be the most important executioner caspases. Two other caspases, Decay and Damm, may be executioner caspases, based on their sequences, but this is speculation.

In general, executioner caspases identified in other animals have essentially the same specificities as executioner caspases in vertebrates. In the discussion that follows, we focus on substrates of mammalian executioner caspases, which are the best characterized.

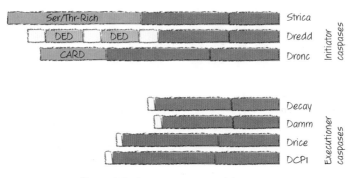

Figure 2.5. Caspases in *Drosophila*.

CASPASES SHOW PREFERENCES FOR THEIR SUBSTRATES

Caspase-mediated cleavage events can have catastrophic consequences that ensure not only that the cell dies, but that it does so quickly and cleanly. To understand this, we must discuss the sequences that executioner caspases recognize in proteins. And for this, we need to become familiar with a bit of terminology.

We know that most caspases prefer to cut their substrate proteins just after aspartic acid residues, and this target amino acid after which the cut occurs is called the P1 residue of the substrate. The amino acid that follows the cut is called P1'. The residue just before P1 is P2, preceded by P3, P4, and so on (Fig. 2.6).

In nearly all caspase substrates, P1 is obviously aspartate, although very rarely it is glutamate. However, the presence of an aspartate in a substrate is not enough to predict that a caspase will cut at that site. First, the aspartate must be exposed so that the enzyme can access it, and second, other amino acids contribute to recognition by the caspase. Based on the use of peptide libraries, caspase-3 and -7 share preferences for sequences that contain DXXD/G (or S or A), where "X" is any amino acid, and "/" indicates the cleavage between P1 and P1'. Other caspases display different preferences, although D is generally preferred at P1. Some of these preferences[2] are shown in Figure 2.7.

How do these preferences apply to actual cellular substrates for caspases? We can identify protein substrates for the caspases by adding active caspases to cell extracts and looking for cleavage events. In this way, hundreds of substrates for caspase-3 have been identified, and we think there may be a thousand or more. Some of these are listed in Table 2.1.

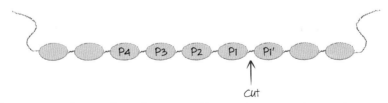

Figure 2.6. Numbering scheme for amino acids around a caspase cleavage site.

---

[2] These preferences are *not* specificities. It is easy to fashion substrates that detect caspase cleavage. When an extract of apoptotic cells is added to such substrates, how much of each peptide substrate is cleaved can be determined, and this information is often extrapolated to define which caspases are active and to what extent. But this approach assumes specificity, a potentially fatal mistake: If a caspase, such as caspase-3, is active and in great excess, then even if it cleaves a peptide relatively poorly, this may be read as activity of a different caspase. Unfortunately, many conclusions have relied on such an approach to determine whether a particular caspase is active under a given condition. Such studies should be treated cautiously.

Caspase-1   WEHD
Caspase-2   VDQQD
Caspase-3   DELD
Caspase-4   LEVD
Caspase-5   (W/L) EHD
Caspase-6   (T/V) QVD
Caspase-7   DEVD
Caspase-8   LETD
Caspase-9   LEHD

Figure 2.7. Caspase preferences, based on peptides. Sequences shown are optimally cleaved, although other peptide sequences may be cleaved nearly as well.

IDENTIFYING CASPASE SPECIFICITY WITH COMBINATORIAL PEPTIDE LIBRARIES

One way to determine specificity is to identify sites in substrates and compare them, as we will see. Another is through the use of combinatorial peptide libraries. This approach requires that we have a method to detect that cleavage of a substrate has occurred. This is most easily done by attaching a fluorescent molecule to the end of a peptide in such a way that it only becomes fluorescent when the bond between it and the peptide is cleaved, resulting in a measurable signal. This is a useful way to measure protease activity in general (Fig. 2.8).

Because for most of the caspases, P1 is D (aspartate), this position can be fixed, and then P1', P1, P2, P3, and P4 (or more) can be varied with every possible amino acid to make a combinatorial library of the set. Each is then tested for its sensitivity to cleavage. An example of the preferences for caspase-3 found in this way are represented in Figure 2.9.

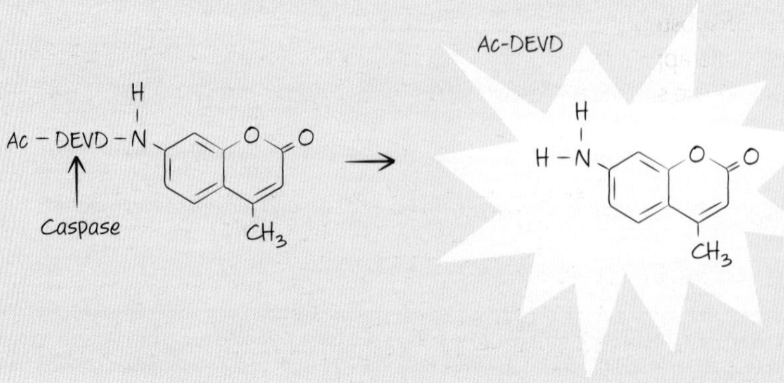

Figure 2.8. Caspase activity detected using a peptide substrate.

(Continued)

(Continued from previous page)

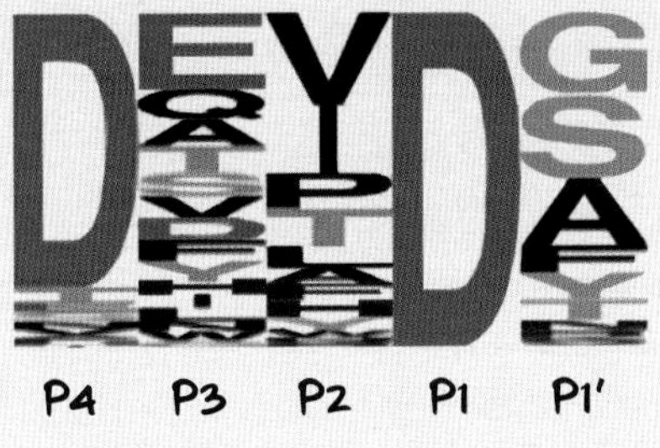

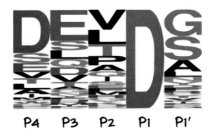

Figure 2.9. Representation of amino acid preferences for caspase-3. Amino acids are shown in single-letter code, and the size of each letter corresponds to its frequency in cleaved peptides.

When all of the cleavage sites in caspase-3 substrates are compared, we get the pattern shown in Figure 2.10.

Clearly, the combinatorial libraries worked well, giving us a pretty good idea of the preferred sites for caspase-3 in native proteins. However, there may be another problem here: When we make a cell extract, we may allow caspase-3 (for example) to access substrates it would not normally see in cells. For example, cells have many subcellular compartments from which caspases may be excluded, and this compartmentalization can be lost when lysing the cell to make an extract. Does caspase-3 have a similar substrate preference during apoptosis in intact cells?

We can address this question by comparing the proteins that are cut during apoptosis. Although we cannot be sure that all of the cuts are due to caspase-3, an exhaustive analysis of substrate cleavage during apoptosis can be performed. The result of such a large-scale approach by mass spectroscopy is informative. This approach identified not only those substrates listed in Table 2.1, but more than 1000 substrates over-

Figure 2.10. Representation of caspase-3 cleavage sites in known substrates.

Table 2.1. List of Some Caspase Substrates

Substrate	Putative or demonstrated functional consequence	Cleavage site (D residue number)
Acinus	Involved in chromatin condensation	DELD (1093)
AKT	Loss of kinase activity. Putative: loss of survival signaling	ECVD (462), TVAD (108), EEMD (119)
β-Catenin	Reduced α-catenin binding. Putative: loss of cell adhesion	YQDD (145), NDED (164), SYLD (32), ADID (83), TQFD (115), YPVD (751)
c-IAP1	Loss of signaling functions	ENAD (372)
E-Cadherin	Release of intracellular fragment	DTRD (750)
GATA-1	Loss of transcriptional activity, which leads to impaired erythropoiesis	EDLD (125)
Gelsolin	Loss of binding to monomeric actin and triggering of F-actin depolymerization, membrane blebbing	DQTD (403)
iCAD	Release of active CAD endonuclease	DETD (117), DAVD (224)
iL-33	Putative: activation of IL-33, which sensitizes toward a TH2 immune response	DGVD (178)
iPLA2	Increased phospholipid turnover, release of lysophosphatidylcholine (LPC), which attracts monocytic cells	DVTD (183)
Lamin A/C	Breakdown of nuclear envelope	VEID (230)
Lamin B1	Nuclear lamia disassembly	VEID (231)
Mst1	Kinase constitutive active; overexpression of amino-terminal fragment induces apoptotic morphology	DEMD (326)
NDUFS1	Disruption of electron transport (complex I), ΔΨm, leading to production of ROS, loss of ATP production, and mitochondrial damage	DVMD (255)
PAK-2	Kinase constitutive active and activates c-Jun amino-terminal kinase pathway; overexpression of cleaved fragment leads to apoptotic morphology (shrinkage and rounding up)	SHVD (212)
PARP-1	Loss of catalytic activity	DEVD (214)
Pro-caspase-3	Activates protease activity	ESMD (28), IETD (175)
Pro-caspase-7	Activates protease activity	DSVD (23), IQAD (198)
RIP-1	Inhibits NF-κB activation	LQLD (324)
ROCK-1	Constitutively activates kinase activity and drives cell contraction and blebbing. Phosphorylates PTEN which then inhibits Akt	DETD (1113)
TRAF-1	Inhibits NF-κB activation, pro-apoptotic	LEVD (163)
Vimentin	Disrupts intermediate filaments, pro-apoptotic	DSVD (85)
XIAP	Amino-terminal fragments inhibit caspase-3 and -7 activity. Carboxy-terminal fragment inhibits caspase-9 activity	SESD (242)

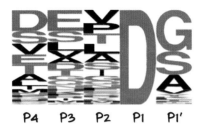

Figure 2.11. Representation of protein cleavage events in cells undergoing apoptosis.

all, and revealed a sequence preference (Fig. 2.11) that matched that predicted by the approaches mentioned above. Caspases with the preference of caspase-3 and -7 may therefore predominate during apoptosis to cut protein substrates. It is not unreasonable to suppose that these are, in fact, caspase-3 and -7.

KEY SUBSTRATES FOR EXECUTIONER CASPASES IN APOPTOSIS

We know what sort of sequences the executioner caspases like to cut during apoptosis, but how does this lead to the features of this form of cell death? Is the effect of executioner caspase activation really "death by a thousand cuts"? Or do only a few cuts in certain key substrates produce the phenomenon we call apoptosis?

Some insights come from considering how caspase cleavage of a substrate can affect a cell. We can envision four ways: Cleavage could (1) destroy an activity, provided that the cleavage is efficient, (2) trigger an activity by removing an inhibitor or an inhibitory domain in the substrate, (3) convert a protein into a dominant-negative version that inhibits the activity of the intact protein, or (4) have no relevant effect at all. The last is vexing and subtle. Many caspase cleavage events may simply "happen" in a cell that is doomed to die anyway. Therefore, even if an event seems likely to be important, upon reflection, its importance may pale in the context of death. For example, the activation of an important transcription factor by an executioner caspase is unlikely to result in the production of a protein if at the same time the genome has been dismantled by DNA fragmentation.

The vast majority of proteins cleaved during apoptosis do not have established roles in the events that characterize this form of cell death. To gain an understanding of how executioner caspases bring about apoptosis, a different but complementary approach is needed.

A useful way to identify caspase substrates that are important for apoptotic events is to dissect a process that depends on caspase activity and identify the key substrate responsible for that process. Below, we consider the substrates whose cleavage produces specific apoptotic events in different cellular compartments: the nucleus, plasma membrane, and mitochondria.

NUCLEAR EVENTS MEDIATED BY CASPASE CLEAVAGE OF SPECIFIC SUBSTRATES

A striking feature of apoptosis is the fragmentation of chromatin. The DNA is cut into pieces equivalent to one or more nucleosomes (multiples of 180 base pairs), and this phenomenon depends on executioner caspases being active in the dying cell.

If we treat normal cytosol from living cells with caspase-3 and then add this to isolated nuclei, DNA fragmentation characteristic of apoptosis occurs. This approach allows identification of the enzyme responsible, as well as elucidation of how it is activated by caspase-3. The enzyme is a nuclease that preferentially cuts DNA at accessible sites between nucleosomes: CAD (caspase-activated DNase) or, alternatively, DFF40 (DNA fragmentation factor of 40 kDa). It is present in healthy cells, but it is held in an inactive complex by an inhibitor, called iCAD (also called DFF45). The inhibitor is the caspase substrate; when cleaved by caspase-3, it releases the active nuclease to cut the DNA.

It would seem especially dangerous for cells to produce a nuclease capable of fragmenting the genome, but there is an additional safeguard. CAD is completely inactive unless properly folded by a chaperone, and this is a function of iCAD, which then holds the nuclease inactive unless it is cleaved. The basic scheme is shown in Figure 2.12.

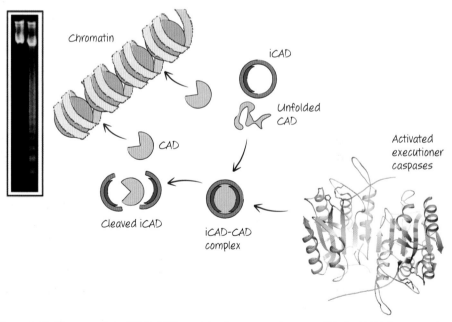

Figure 2.12. Cleavage of the iCAD-CAD complex by caspases is responsible for DNA fragmentation during apoptosis. Caspase-3 cleaves iCAD, releasing the active CAD, which randomly cuts the chromatin at accessible sites between the nucleosomes. When run on an agarose gel, the DNA forms a ladder composed of multiples of nucleosome-sized lengths.

Cutting of the chromatin by CAD probably facilitates the degradation of the dead cell after it has been engulfed by a phagocytic cell (discussed in Chapter 9). In the absence of functional CAD (achieved by removing CAD or its chaperone, iCAD), DNA fragmentation does not occur in the cell before engulfment, but the cell dies nevertheless. Therefore, although it is extremely unlikely that the cell can survive once caspases have cleaved iCAD and released active CAD, this is not required for cell death.

So, are CAD and iCAD important? It is difficult to say. Both are present in most of the animals for which we have genomic sequences, with at least one exception. Among the animals, iCAD and CAD are found in hydra, sea anemones, insects, and of course vertebrates (to name a few), but they are not found in nematodes. Apparently, nematodes lost this substrate and its function in apoptosis along the way.

Other nuclear events in apoptosis can be traced to the cleavage of additional caspase substrates. During apoptosis, the chromatin becomes condensed, and this has been linked to the cleavage of the protein Acinus by caspase-3 (and probably caspase-7). Precisely how Acinus causes chromatin condensation is unclear but may involve phosphorylation of histones or its participation in DNA fragmentation by CAD.

Disruption of the nuclear envelope also occurs in apoptosis, and this can be traced to the cleavage of lamins by caspase-6. Lamins are important structural elements that preserve nuclear integrity and are involved in the breakdown and regeneration of the nuclear envelope during mitosis. If mutant lamins lacking the cleavage site are introduced, the nuclear envelope remains intact during apoptosis; however, all other aspects of apoptosis seem to proceed normally.

EVENTS AT THE PLASMA MEMBRANE CAUSED BY CASPASES

During apoptosis, the cell undergoes extensive membrane blebbing, extending bulbous outgrowths that often break off as small, membrane-bound bodies. This is an effect, at least in part, of actin polymerization, and pharmacological inhibitors of actin polymerization block blebbing.

Three caspase substrates have been implicated in actin polymerization and blebbing during apoptosis. These are the actin regulator gelsolin and two kinases that function in signaling pathways that control actin organization in cells: p21-activated kinase (PAK) and ROCK-1 kinase. In all three, caspase cleavage activates the protein by removing regulatory domains, which results in dynamic changes in actin organization to cause blebbing. Pharmacological inhibitors of ROCK-1 prevent blebbing and can even stop it after it is under way. As with the other related events that we have discussed, however, the cell still dies.

One important consequence of apoptotic cell death is that the cell gets eaten by other cells before the integrity of the plasma membrane is lost. The "eat me" signals that ensure that this will occur depend on caspase activation. The most important of these signals is the appearance on the cell surface of the lipid phosphatidylserine, which is recognized by the cells that do the eating (discussed in much more detail in Chapter 9). Phosphatidylserine is normally restricted to the inner leaflet of the plasma membrane, but during apoptosis, the lipids scramble, and phosphatidylserine is exposed. Unfortunately, although this is perhaps one of the most important events during apoptosis, we have yet to identify the caspase substrates that are responsible for phospholipid scrambling and phosphatidylserine externalization.[3]

MITOCHONDRIAL EFFECTS OF EXECUTIONER CASPASES

When the mitochondrial pathway of apoptosis is engaged (discussed in detail in Chapter 4), activated executioner caspases gain access to the inner membrane of mitochondria (because the outer mitochondrial membrane becomes permeable). Here, several caspase substrates on the inner membrane are found, including NDUFS1, which is an integral part of complex I of the electron transport chain, the major source of energy in the cell.

When NDUFS1 is cut by caspases, several events rapidly follow. Electrons that would normally be used for oxidative phosphorylation are shuttled to oxygen to produce superoxides. These can then be converted to hydroxyl radicals that are damaging to many proteins and membranes. In addition, the proton gradient normally produced by electron transport dissipates, and ATP levels rapidly fall. This loss of ATP has effects on the plasma membrane, because ion pumps require ATP to operate. Introduction of a noncleavable NDUFS1 mutant into cells prevents these rapid, caspase-dependent events, but has no effect on other apoptotic changes, such as DNA fragmentation and blebbing.

DEATH BY A THOUSAND CUTS?

As we noted, there are many hundreds of caspase substrates that are cleaved when executioner caspases become active. How many of these are actually responsible for cell death? Clearly, if a nuclease such as CAD/DFF45 becomes active in a cell and fragments the DNA extensively, the cell will not survive. But some specialized cells, such as mammalian red blood cells, persist for more than 100 days without a nucleus (the elimination of the nucleus in red blood cells does not depend on CAD). And cells lacking this caspase substrate still die by apoptosis. A glance at the substrates in Table 1 and those we discussed above give us an idea why. Indeed, it is remarkable that a few key substrates that have major roles in producing specific changes that we see during apoptosis can be identified. But it is unlikely that only a few key substrates (these or others) are responsible for cell death.[4]

---

[3] Some readers may find this an exciting opportunity.

[4] There are many reports that a favorite substrate is paramount in cell death. But in every case, introduction of high levels of such proteins (cleavable or mutated to avoid cleavage) may have other effects that alter the biology of the cell and its sensitivity to a given apoptotic challenge, and we should ask (as these investigators apparently have not) what happens to all of the other substrates and their functions when caspases are active.

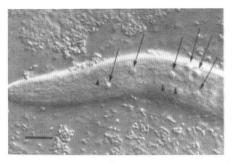 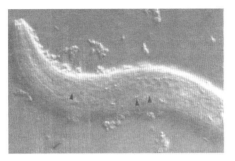

Figure 2.13. Extra cells in CED3 mutant nematodes. Normal cell deaths in a wild-type larva (arrows, *left*) are not seen in a CED3 mutant animal (*right*).

CASPASES AS KILLERS

The short answer to the question of what kills the cell is, of course, "caspases." Introduction of an executioner caspase into any type of cell causes it to die, and in animal cells, this occurs by apoptosis.

As we go into the pathways of apoptosis in this book, we will see several examples of where the activation of caspases is not essential for cell death, and other mechanisms come into play. For now, it is important to be aware of studies that reveal just how important caspases really are. For example, in *C. elegans*, one caspase, CED3, is required for most cell death during development and in germ cells of the adult worm that die in response to stress. Animals with mutations that prevent CED3 activation or function accumulate extra cells that differentiate and persist (Figure 2.13). In this animal, cell death depends on completion of the apoptotic pathway and therefore depends on the caspase.

This is also true in *Drosophila*, although again, not all cell deaths follow this rule. *Drosophila* lacking the initiator caspase Dronc accumulate extra cells, which is lethal for the developing fly (Fig. 2.14).

Figure 2.14. Extra cells in Dronc-deficient fly embryos. Cell death (stained blue) in wild-type embryos (*left*) are not seen in embryos lacking Dronc (*right*).

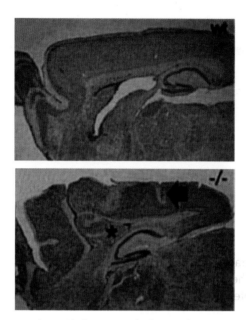

Figure 2.15. Extra cells (arrow and asterisk, *bottom*) in the brains of caspase-3-deficient mice (*bottom*) compared to wild-type mice (*top*).

Cultured *Drosophila* cells that lack caspase activity continue to survive and proliferate when exposed to stresses that would normally cause apoptotic cell death, emphasizing the importance of these enzymes in cell demise. Mice lacking the initiator caspase, caspase-9, or the executioner caspase, caspase-3, also accumulate extra cells (Fig. 2.15). Elimination of caspase-7 as well as caspase-3 exacerbates this effect.

Caspases must therefore be important for death of some cells. However, a close look at the development of such mutant mice reveals that developmental cell death still occurs. We return to this problem of caspase-independent cell death later (e.g., in Chapter 8) as we learn more about the pathways and other forms of cell death.

So far, we have discussed caspases and the way they contribute to apoptosis. But how are the caspases activated to cause this devastation? We consider this problem next.

CHAPTER 3

Caspase Activation and Inhibition

ACTIVATION OF EXECUTIONER CASPASES

Executioner caspases, when activated, cleave hundreds or thousands of substrates in the cell to orchestrate apoptosis. In most animals (the exception, as we have noted, is in nematodes), the executioner caspases have short prodomains lacking interaction sites for other proteins. The inactive proforms of these caspases exist in the cell as dimers. In this form, they are constrained from forming an active site until they are cleaved between the large and small subunits. This cleavage permits the chain–chain interaction that snaps the two active sites into place, allowing the now mature protease to be maximally functional. This is an important rule for understanding apoptosis that bears repeating:

> The executioner caspases that orchestrate apoptosis preexist in cells as inactive dimers that are activated by cleavage between the large and small subunits.

In the inactive proforms of the executioner caspases, the catalytic diads are not in position to gain access to the target aspartate in the substrate protein. A look at one procaspase structure shows us why (Fig. 3.1, left). This is the structure of inactive procaspase-7, and we can see how it changes as it becomes activated (Fig. 3.1, right).

Note that both the inactive and active forms of the enzyme are dimers (remember the rule above), with two active sites. On activation, the active cysteine-histidine diads do not move; what changes is the structure of the loop that ultimately forms the substrate specificity pocket (the arginine that interacts with the target aspartate in the substrate is indicated). The reason that this shape change occurs on cleavage is shown in Figure 3.2.

In the inactive form, the center of the dimer is occupied by the region of the caspase that serves as the linker between the large and small subunits. When the linker is cleaved, it leaves the central region and each end interacts with an end of the cleaved linker from the other chain, stabilizing the structure and holding the ends away from the center. Now, each loop comprising the specificity pockets can snap

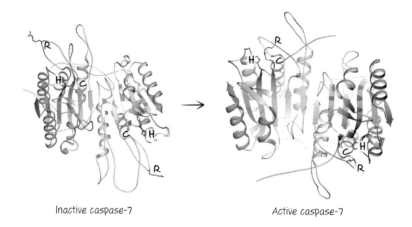

Figure 3.1. The structures of inactive and active caspase-7. The arginine (R) in the specificity loop is indicated, as is the cysteine (C)-histidine (H) catalytic diad.

into position, drawn into place by a neighboring loop. The "elbow" of the specificity loop stretches into the center of the dimer, the region previously blocked by the linker. The result is the formation of two active sites in the mature caspase.

Once activated in this way, the caspases act in *trans* to remove the small prodomains as well. This changes the size of the mature protein, but it does not seem to be important for executioner caspase activation or function.

What mediates the cleavage event that activates the procaspases? In the case of caspase-6 in mammals, the cleavage responsible for activation is by caspase-3 and -7. To

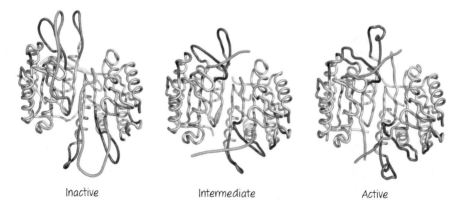

Figure 3.2. Close-up view of caspase-7 activation. Three loops mediate caspase-7 active site assembly: the specificity loop (red), a neighboring loop that draws the specificity loop into position (blue), and the linker between the large and small subunits (green). Structures are the inactive zymogen (*left*), the inhibitor-bound enzyme (*center*), and the active enzyme (*right*).

some extent, caspase-3 and -7 can be activated by other active molecules of caspase-3 and/or -7.

But this begs the question: What first cleaves and thereby activates a few executioner caspases to cause apoptosis? The answer, in the vast majority of cases, is the initiator caspases, in particular, caspase-8 and -9 in vertebrate cells. However, before considering these caspases and how they work, we briefly consider another enzyme capable of cleaving and activating the executioner caspases in vertebrate cells. In doing so, we delineate our first, albeit simplified, apoptotic pathway.

INTERLUDE: GRANZYME B AND APOPTOSIS INDUCED BY CYTOTOXIC LYMPHOCYTES

Viruses are tricky little things. They quickly infect and subjugate cells for the production of more virus, posing special problems for multicellular organisms seeking to eliminate these pests. In vertebrates (and perhaps other animals), one approach to this problem is cytotoxic lymphocytes (cytotoxic T cells and natural killer cells) that identify infected cells and cause them to undergo apoptosis, destroying both them and the parasites that they harbor. Figure 3.3 shows the killing of a target cell by a cytotoxic lymphocyte. The cell death has the characteristics of apoptosis.

Cytotoxic lymphocytes have a number of means at their disposal for murdering their target cells, but here, we concern ourselves with only one—the function of cytotoxic granules—and only one of the ways in which these work. During cell killing, the contents of the cytotoxic granules are released onto the membranes of the target cells. The released molecules include a pore-forming protein, perforin, and a collection of proteases called granzymes. One of the latter is granzyme B. Granzyme B is a protease, but unlike caspases, it is a serine protease (i.e., its active site contains a serine, rather than a cysteine). However, like the caspases, granzyme B is an endopeptidase that cleaves after aspartic acid residues, such as those involved in activating executioner caspases.

When the cytotoxic lymphocyte contacts a target cell and releases its granule contents, perforin permits the entry of the granzymes into the cell that is the victim of

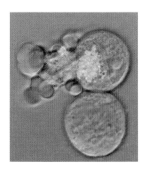

Figure 3.3. Cytotoxic lymphocyte killing a target (cell). The target cells are stained red, and the cell at *left* is undergoing apoptosis. (Green) Cytoxic granules in the cytotoxic lymphocyte.

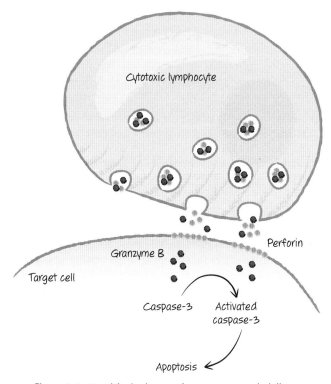

Figure 3.4. Simplified scheme of cytotoxic granule killing.

this murderous assault (Fig. 3.4).[1] If sufficient granzyme B is present, it cleaves caspase-3 and -7 at the aspartate between the large and small subunits of each, activating the caspases. The result is a feed-forward activation that kills the cell by apoptosis, effectively removing the compromised cell and the virus in the process.

This mechanism is not restricted to viral infections; it includes immune responses by cytotoxic lymphocytes to other intracellular infections as well as responses to tumor cells and foreign tissue grafts. This mechanism of killing is not the only way in which cytotoxic lymphocytes work, nor is it the only way granzyme B can function to trigger apoptosis. However, it illustrates how caspase activation can occur in this bona fide apoptotic pathway.

ACTIVATION OF INITIATOR CASPASES

In the vast majority of cases in which apoptosis occurs, the proteases that trigger executioner caspase activation are the initiator caspases. In mammals, these are pre-

---

[1] It is probably not as simple as this. Granzyme B is taken up by target cells, even without perforin, but perforin is necessary for the granzyme to gain access to the cytoplasm.

dominantly caspase-8 and -9. When activated, they can cleave and thereby activate the executioner caspases, resulting in apoptosis.

Unlike executioner caspases, initiator caspases (as well as other caspases, such as inflammatory caspases and caspase-2) exist in cells as inactive monomers. And also unlike executioner caspases, these are not activated by cleavage, but instead, by dimerization. This is a second key rule of caspase activation that bears repeating:

> Initiator caspases preexist in cells as inactive monomers. They can only be activated by dimerization.

Once the initiator caspases dimerize, they become active and can cleave themselves between the large and small subunits, which acts to stabilize the dimer. Therefore, although cleavage does not activate the caspase, it does have a function.

The idea that the activation of initiator caspases involves bringing monomers together to form dimers is called the "induced proximity" model. The principle can be shown experimentally using recombinant caspases that lack the prodomains. For example, when such a truncated caspase-2, -8, or -9 is produced, each is enzymatically inactive. But if salt conditions are altered to promote aggregate formation, caspase activity quickly appears. This happens regardless of whether the caspases can undergo cleavage, that is, even if the cleavage sites between the large and small subunits have been mutated (Fig. 3.5).

When the salt is removed, the cleaved active dimers remain active. But if the caspase was uncleavable (because of mutation), its activity is rapidly lost. This is why we

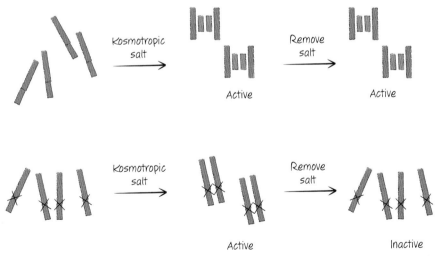

Figure 3.5. Initiator caspases can be activated by induced proximity. Kosmotropic salts, which cause protein aggregation, activate initiator caspases that remain active after the salt is removed. If the cleavage sites between the protease subunits are made uncleavable by mutation, the salts can still activate the enzyme, but the enzyme becomes inactive when the salts are removed.

say that cleavage of initiator and related caspases is not required for activity but serves to stabilize the dimers.

This stability, however, is important. Although a noncleavable mutant of caspase-8, for example, can be demonstrably activated by dimerization, it does not promote apoptosis. We return to this when we consider the biological functions of caspase-8 in apoptosis and other phenomena. But now, there arises a major question: What dimerizes the initiator caspases (and the other caspases that behave in this way)? The answer is adapter proteins.

ADAPTER PROTEINS INTERACT WITH CASPASES VIA DEATH FOLDS

All caspases that are not executioner caspases have long prodomains containing regions that interact with other proteins, the adapter proteins. To activate caspases, however, these adapter proteins must not only bind the caspases but also dimerize them. In general, this occurs through processes that dimerize or oligomerize the adapters. How this happens is through distinct interactions, depending on the adapter. These interactions, the adapters involved, and the initiator caspase engaged all define the different apoptotic (or related) pathways that are the subjects of the following chapters. All of these apoptotic pathways, however, have the basic form shown in Figure 3.6.

The protein–protein interaction regions in the prodomains of initiator, inflammatory, or other caspases are of different types, depending on the caspase. Caspase-8

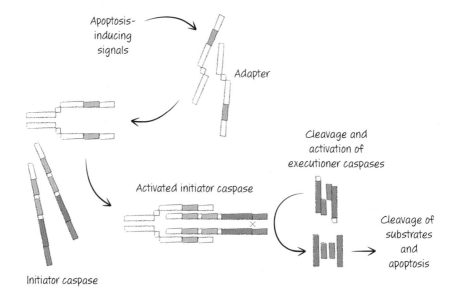

Figure 3.6. General apoptotic pathways.

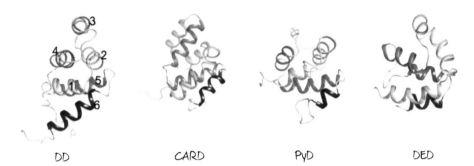

Figure 3.7. Representative death folds. Although different death folds do not have sequence similarity, they are structurally related. Each has six globular helical bundles, shown here colored gray to dark blue from the amino to carboxyl termini.[2]

and -10 have two death effector domains (DEDs). Caspase-1, -4, and -5 (and in rodents, caspase-11), caspase-2, and caspase-9 (the initiator caspase) all have caspase-recruitment domains (CARDs). DEDs and CARDs are also involved in other protein–protein interactions not involving caspases, so the presence of such a domain is not itself a demonstration that the protein is involved in caspase activation or apoptosis.

Although DEDs and CARDs are not related by sequence homology, they are structurally similar (Fig. 3.7). This type of structure is called a "death fold." Other sequences that form death folds are also shown in Figure 3.7, including a death domain (DD) and a pyrin domain (PyD). Although these are not generally found in mammalian caspases, they are found in some of the molecules leading to caspase activation. As an intriguing aside, one of the caspases identified in zebrafish has a PyD in its prodomain.

In general, these protein–protein interaction regions work by like–like recognition. That is, a protein that binds to a CARD in a caspase will do so via its own CARD. Although not every CARD (DED, DD, or PyD) will interact with every similar domain (in fact, it generally interacts with only one other protein, although there are exceptions), the binding will be via like–like domain recognition.

This theme frequently recurs in the chapters that follow, with sentences that take on the form "the X domain in protein A binds to the X domain of protein B."

Why things have turned out this way is not at all obvious. It does, however, seem as though there is a deep but elusive evolutionary principle at work in these interactions.

---

[2] The molecules from which these particular structures were obtained are FADD (DD and DED), APAF1 (CARD), and NALP1 (PyD). These molecules are considered in later chapters, but for now the main point is that the different death folds are structurally related.

ACTIVATION OF INITIATOR CASPASES BY DIMERIZATION

How does dimerization activate initiator caspases, and why is cleavage not required for proteolytic function but only for stabilization of the dimer? The proforms of the initiator (and other) caspases, as with the executioner caspases, are inactive because the specificity-determining pocket in the enzyme is disorganized and out of position to bring the substrate to the catalytic cysteine-histidine diad. However, unlike the case in executioner caspases, in initiator caspases this has nothing to do with the linker region between the large and small subunits, which is not cleaved. Instead, another loop from the opposite chain in the dimer is needed to effectively pull the loop containing the substrate-specificity pocket into position. In caspase-9, this event can only happen at one protease site at a time, due to constraints in the dimer. Figure 3.8 shows this interaction in the caspase-9 dimer that produces an active site.

Clearly, this formation of an active site in the protein depends on the formation of the dimer. When the inactive procaspase monomers are brought together by the appropriate adapter protein, the interaction results in the formation of an active site.

This has practical consequences for our understanding of apoptotic pathways, because although initator caspases undergo cleavage upon activation, cleavage does not itself activate them. Unfortunately, many studies have relied on detection of cleaved initiator caspases as an indication that the caspase has been activated en route to apoptosis. Such studies must be treated with caution. Similarly, a misunderstanding of the role of cleavage in initiator caspase activation has led some investigators to suggest that proteases that cleave initiator caspases thereby activate them, but we can now see that cleavage of the inactive monomer cannot do so. As with many areas of science, the fact that a conclusion is published does not make it true.

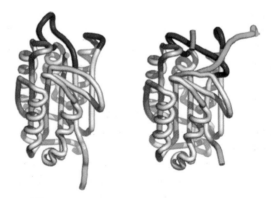

Figure 3.8. The transition from inactive to active caspase-9. When caspase-9 is dimerized, the specificity-determining loop (red) is brought into position by a loop (blue) from the other monomer. Cleavage of the region between the protease subunits (green) is not necessary for activation. Only one site in the dimer can form at a time: One is inactive (*left*) and the other is active (*right*).

INITIATOR CASPASES CAN BE ACTIVATED BY DIMERIZATION WITH OTHER MOLECULES

The way in which initiator caspases are activated makes it possible for other proteins to induce the conformational changes necessary to form an active site in the caspase. Two such proteins are FLIP and MALT, both of which can be brought by adapter proteins into heterodimers with the initiator caspase, caspase-8.

FLIP is closely related to caspase-8 at the sequence level but lacks the cysteine necessary for proteolytic activity. When dimerized with caspase-8, FLIP interacts with the caspase so that it forms an active site, but because FLIP is not a protease, caspase-8 is not cleaved. When FLIP is present, it prevents caspase-8–mediated apoptosis. This dimer, however, is important for a signaling event that is needed for cell survival in certain circumstances (see Chapter 9).

MALT is not related to caspases at the sequence level, but it has predicted structural similarities that classify it as a paracaspase.[3] Like FLIP, MALT can be brought by adapter proteins into heterodimers with caspase-8. Again, the activated caspase-8 is not cleaved, and the function of this heterodimer remains obscure. Nevertheless, it illustrates the principle that caspases need not necessarily be *homodimers* to be active.

OTHER INITIATORS AND EXECUTIONERS IN THE CASPASE COLLECTION

So far, we have focused on mammalian caspases in our consideration of how they work. But as we know, apoptosis is not restricted to vertebrates. A brief tour of the caspases in other organisms is warranted.

Two initiator caspases have been identified in *Drosophila*. Dredd, which contains DED regions in its prodomain, appears to have a nonapoptotic function that remains elusive. The other, Dronc, which has a CARD-containing prodomain, clearly functions as an initiator caspase, activating the executioner caspases Drice and DCP1 by cleaving them. Like other initiator caspases, Dronc is activated by dimerization. Another caspase, Strica, seems to have an unusual prodomain and may be an initiator caspase, although this is far from proven.

In *Caenorhabditis elegans*, only one caspase is known to function in apoptosis. This caspase, CED3, has a long prodomain containing a CARD and appears to be activated by dimerization. Once dimerized, CED3 then functions as an executioner caspase, causing the cell to die.

Is the process in *C. elegans* representative of the primordial apoptotic pathway, where activating one caspase causes death? Perhaps not. Cnidaria (including the hydroids, jellyfish, and corals), which are probably older in evolutionary terms than the

---

[3] See Footnote 1, Chapter 2.

nematodes, have both initiator- and executioner-type caspases, based on sequence. This might suggest that somewhere along the way, nematodes lost caspases from the genome. This is not unlikely; the purple sea urchin (*Strongylocentrotus purpuratus*), an echinoderm, has 31 different caspases; compare that with the relatively small number we find in vertebrate species. In time, we hope to learn how these novel caspases function in cell death or other phenomena in such organisms.

INFLAMMATORY CASPASES FUNCTION IN SECRETION AND CELL DEATH

In mammals (and other vertebrates), one caspase has a well-characterized role that is distinct from induction of cell death (although, as we will see, it *can* cause cell death). This is caspase-1. Like initiator caspases, it exists in cells as an inactive monomer and is activated by dimerization. However, at low levels of the active caspase, caspase-1 functions not as a killer, but rather, in inflammation. This is because caspase-1 processes (by cleavage) the precursor forms of two related cytokines: interleukin-1β and interleukin-18. These cytokines have important roles in inflammatory responses. The pathways involved in caspase-1 activation are discussed in Chapter 7, but for now, it is worth noting that the same principles we have learned that apply to apoptosis apply to other caspases as well, even if the result is not apoptosis.

Caspase-1 occupies a position in the mammalian genome in close proximity to other related caspases, including caspase-4, -5, and -12 in humans and caspase-11 and -12 in rodents. For this reason, these are often also referred to as inflammatory caspases. In some cases, caspase-5 (human) and caspase-11 (rodent) may participate in caspase-1 activation, as we will see later, but caspase-4 and -12 do not. The function of caspase-4 remains fairly mysterious.

Caspase-12 is particularly intriguing. It is found throughout mammals and primates, until we examine humans. Among sub-Saharan Africans, caspase-12 is expressed in about 20% of individuals. In contrast, most humans have a stop codon in the third exon that results in an unstable truncated protein. Evidence suggests that this null allele was positively selected during the African migration about 67,000 years ago. To get an idea of why this might have occurred, we have to consider what caspase-12 does. In general, the caspase itself appears to have very little activity. When coexpressed with caspase-1, it dampens the activity of the latter. It may be that under some conditions where strong caspase-1 function is a problem, this dampening function is favored, whereas under other conditions, increased caspase-1 activity provides protection (e.g., from some infections). In any case, it *may* be correct to think about caspase-12 more as a regulator of protease activity than as a protease per se.

Caspase-1, in addition to processing interleukin-18 and interleukin-1β, appears to have another role in inflammation that is not well understood, functioning in an unusual mode of secretion in the cells that express it. This secretion does not involve signal peptides or the endoplasmic reticulum (as conventional secretion does), but beyond that,

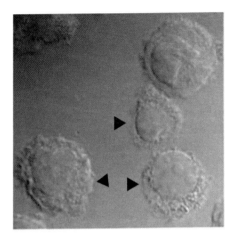

Figure 3.9. Pyroptosis. Cells die in a caspase-1–dependent manner, indicated by arrows.

we do not yet know how such secretion occurs. Proteins that are secreted by this mechanism include interleukin-1β and interleukin-18, and proteins that are not processed by caspase-1, such as interleukin-1α and peptides with antibacterial activities.

Under conditions of intense activation, caspase-1 can trigger cell death. Like initiator caspases, caspase-1 can cleave and activate at least one executioner caspase (caspase-7). But in addition, it may itself cleave other substrates in the cell, resulting in a form of cell death that resembles apoptosis only in part. This caspase-1–triggered cell death is often referred to as "pyroptosis" (Fig. 3.9).

In *Drosophila*, the initiator-like caspase Dredd does not seem to function in cell death but is involved in the secretion of peptides that have antibacterial activity. For this reason, it is therefore tempting to think of Dredd as a sort of inflammatory caspase.

Caspase-2: The "Orphan" Caspase

Caspase-2 was the second mammalian caspase discovered, and by sequence, it is the most highly conserved among animals. Like the initiator and inflammatory caspases, it is activated by induced proximity of the inactive monomers, and cleavage serves to stabilize the mature enzyme. However, unlike the initiator caspases and caspase-1, caspase-2 does not seem to be very good at cleaving and activating the executioner caspases. It can cause apoptosis indirectly by engaging the mitochondrial pathway (discussed in Chapter 7), but this does not seem to be a primary function for this enzyme. We just do not know its primary function.

Caspase-2 has been implicated in the response to a number of stressors, but in no case is it absolutely required for apoptosis. What caspase-2 does and what it is "for" is unknown at present. Is caspase-2 important? In a mouse model of lymphoma, the absence of caspase-2 profoundly accelerates the appearance of tumors. No other caspase has such a profound effect in this model.

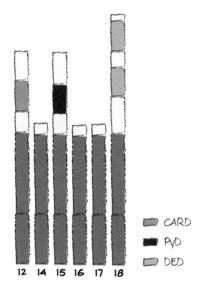

Figure 3.10. Caspase-12 and beyond in mammals. Lengths shown are arbitrary. Caspase-15, -17, and -18 are not found in humans and caspase-12 is frequently not found in humans. Caspase-11 (mouse) and -13 (cow) are homologs of caspase-4 and are not shown.

ON BEYOND 12

Vertebrates have a number of other caspases that have been characterized, and some have multiple copies of the ones that we have discussed. In most cases, their functions are unknown as are the adapter molecules that activate them.[4] Figure 3.10 shows a list of these vertebrate caspases.[5] The figure is provided more as a point of interest and for the reader to get a feeling of the evolutionary plasticity in the system than as important information for the discussions that follow. That said, we may want to ask ourselves why a set of molecules that are so well conserved among the animals show so much variability, duplication, deletion, and modification, if the only role is to orchestrate apoptotic cell death.

INHIBITION OF CASPASES IN CELLS

The importance of caspases in cell death and apoptosis leads us to the question of how they are regulated in cells, because it is almost axiomatic that anything so devastating must be inhibited at several levels. As we will see, however, endogenous caspase inhibitors are essential in some cases, but not all.

---

[4] There is evidence that caspase-14 has a role in the differentiation of skin.

[5] Caspase-13 is not included, because this turned out to be the bovine homolog of caspase-4.

SYNTHETIC CASPASE INHIBITORS

Peptides that are recognized by caspases can be fashioned into caspase inhibitors if a reactive "warhead" is appended to the P1 aspartate that binds (reversibly or irreversibly) to the active cysteine in the caspase. Again, as with the substrates, such inhibitors are not specific, although they are often effective as general caspase inhibitors (that may also inhibit other proteases as well). Therefore, although these can be useful for research, any conclusions that come from using them have to be regarded with care. Examples of synthetic inhibitors based on peptides are shown in Figure 3.11. The structure of caspase-3 bound to a peptide inhibitor is shown in Figure 3.12.

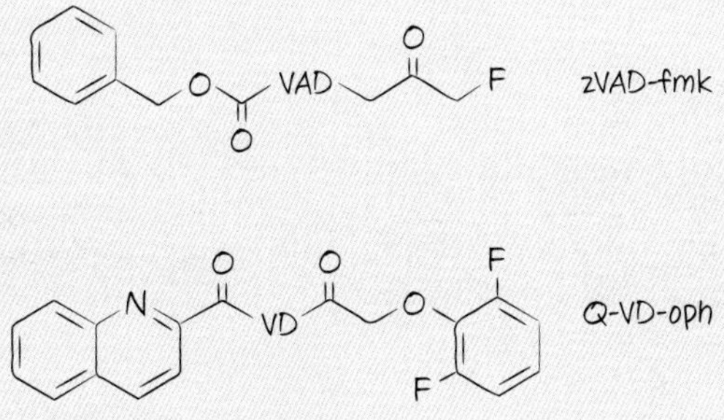

Figure 3.11. Caspase inhibitors. The peptide portion is shown with a single-letter code.

Other inhibitors of caspases that are not peptide based have been developed. These work by either binding to the active site or inhibiting the conformational changes needed for caspase activation.

If we induce apoptosis, for example, by stressing cells, and add agents that inhibit the executioner caspases, all of the characteristic morphological features of apoptosis are blocked (Fig. 3.13).

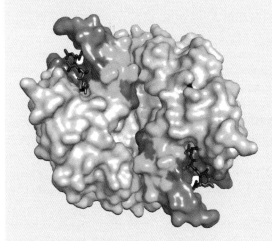

Figure 3.12. Caspase-3 bound to zVAD-fmk.

(Continued)

(Continued from previous page)

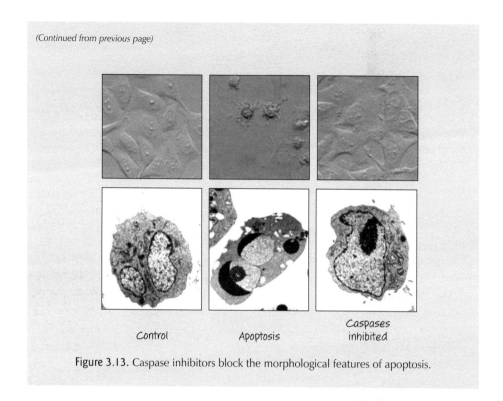

Figure 3.13. Caspase inhibitors block the morphological features of apoptosis.

VIRAL AND ENDOGENOUS CASPASE INHIBITORS: IAP PROTEINS

When a virus infects a cell, it uses its host to make more virus, and if the cell can undergo apoptosis before more virus is made, the infection can be stopped. But viruses have evolved strategies for preventing apoptosis. One strategy is to inhibit caspases.

The insect virus baculovirus actually makes two different caspase inhibitors that work by different mechanisms. One is an inhibitor of apoptosis protein (IAP). When this was first identified, it quickly became apparent that insects have their own IAP proteins ("inhibitor of apoptosis protein proteins" is, of course, redundant, but this phrase persists, and we will not buck the trend here). The most important of these is called DIAP1 (*Drosophila* IAP1, so named before a moratorium was called on *Drosophila* proteins starting with a "D"). DIAP1 binds to and inhibits Dronc and the executioner caspases. Removal of DIAP1 is sufficient to cause apoptosis in fly cells and embryos (Fig. 3.14).

It was also quickly realized that IAP proteins are found in many animals, including humans, based on sequence similarities. All IAP proteins share one or more copies of a motif called the baculovirus IAP repeat (BIR). *But*, and this is very important, despite

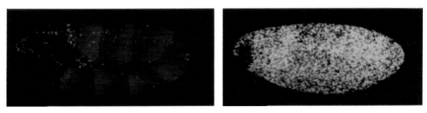

Figure 3.14. Loss of DIAP1 causes apoptosis. Extensive cell death, staining green, is not seen in a wild-type embryo (*left*), but is observed in a DIAP1 mutant embryo (*right*).

their somewhat unfortunate name, *most* IAP proteins do not inhibit caspases or block apoptosis. Some IAP proteins are listed in Figure 3.15. Those capable of directly inhibiting caspases are noted.

It should be immediately apparent that of the mammalian IAP proteins, only XIAP is a direct inhibitor of caspases.[6] In fact, different regions of the protein inhibit different caspases: BIR3 (and the region adjacent to it, the RING, see Fig. 3.15) inhibits the initiator caspase-9, whereas BIR2 (and the linker between it and BIR1) inhibits the executioner caspase-3 and -7. Figure 3.16 shows the binding of XIAP to caspase-3.

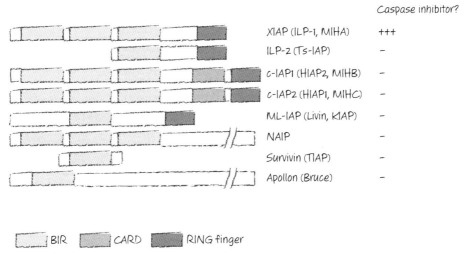

Figure 3.15. Some IAPs.

---

[6] Some studies suggest that cIAP1, cIAP2, NAIP, and survivin also inhibit caspases. All can associate with some caspases (directly or indirectly) and can ubiquitinate them, which may influence activity and/or turnover. However, none of these is a direct inhibitor of caspases, and any inhibition by these other mechanisms appears to be minor.

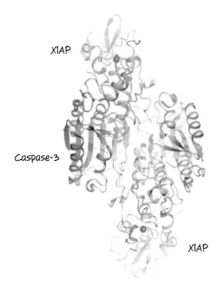

Figure 3.16. XIAP binds to the active sites in caspase-3. Only the BIR2 domain in XIAP (green) is shown. (Purple circles) Zinc atoms that coordinate the folding of XIAP.

In addition to BIR domains, many IAP proteins have another domain called RING that functions as an E3-ubiquitin ligase, responsible for putting ubiquitin chains on target proteins and targeting them for degradation by the proteasome or changing their signaling properties. XIAP and DIAP1 cause degradation of caspases by this mechanism.

Is XIAP essential? That is, if it is absent, do caspases spontaneously activate? Mice lacking XIAP have no developmental defects, so the answer appears to be no. However, we will see later (in Chapter 6) an example of XIAP function in apoptosis.

So, we have a mystery: DIAP1 is essential in the control of *Drosophila* apoptosis, but IAP proteins as inhibitors of caspases do not appear to be crucial in other organisms. Nematodes lack caspase-inhibitory IAP proteins altogether, and XIAP appears, at first pass, to be dispensable, at least in mice. It may be that the control of apoptosis upstream of initiator caspase activation is so tight that control by IAP proteins in most animals is unnecessary. As we go further into the different apoptotic pathways, we will see why this may be so. Alternatively, there may be other caspase inhibitors that we have not sufficiently appreciated.

At this point, we can begin to piece together what we have discussed so far in terms of caspase activation, inhibition, and apoptosis in animals (see Fig. 3.17).

OTHER VIRAL INHIBITORS OF CASPASES

Viruses that infect mammalian cells sometimes carry caspase inhibitors, but these are distinct from the IAP proteins. Pox viruses, for example, express inhibitors belonging

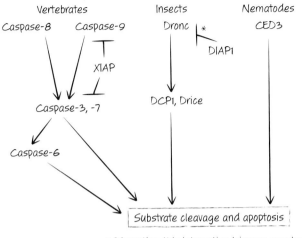

Figure 3.17. Caspases and IAPs in apoptosis.

to the serpin family. Most serpins act as inhibitors of serine proteases (remember that caspases are *cysteine* proteases), but some viral serpins instead act as caspase inhibitors. For example, one serpin, CrmA, expressed by cowpox virus, does not inhibit caspase-9 or the executioner caspases in cells, but it is effective in blocking inflammatory caspase-1 and initiator caspase-8, and it also inhibits the serine protease granzyme B (recall that granzyme B is made by cytotoxic lymphocytes and triggers apoptosis in target cells). As we will see, caspase-8 has roles in some immune effector mechanisms, and therefore this inhibitor may function in immune evasion.

In discussing baculovirus, we mentioned that there are actually two different caspase inhibitors that it expresses. In addition to its IAP, baculovirus also makes p35, an inhibitor not found in any form in animals. This protein is remarkably specific for caspases and acts as a caspase substrate. When it is cut, it becomes irreversibly bound to the caspase. Although unique to insect viruses, p35 can effectively inhibit mammalian (and other) caspases, making it useful for experimental purposes.

We have reached the point in our discussion where we know how caspases work, what they cut, and how they are inhibited. And to a first approximation, we know that apoptosis is caused by key substrates being cleaved by executioner caspases that are themselves activated when they are cut by initiator caspases. The latter are activated when adapter proteins bring them together so that they activate themselves by induced proximity. But what are these adapters and how do they come to engage the initiator caspases? The answer is apoptotic pathways. We will begin with the pathway responsible for most apoptosis in humans (and vertebrates in general): the mitochondrial pathway.

CHAPTER 4

The Mitochondrial Pathway of Apoptosis, Part I
MOMP and Beyond

MITOCHONDRIA AND CELL DEATH

The major mode of apoptosis in vertebrates is the *mitochondrial pathway*. This pathway of cell death is engaged by a vast array of cell stresses, including growth factor deprivation, cytoskeletal disruption, DNA damage, accumulation of unfolded proteins, hypoxia, and many others. It is also activated by developmental signals such as hormones that instruct cells to die. In this chapter, we discuss how mitochondria are involved in the activation of caspases in this type of cell death.

A JUST SO STORY

Before examining the remarkable way in which mitochondria affect apoptosis, it might be worth considering how this organelle may have become involved in cell death—after all, in nearly all eukaryotes, mitochondria are essential for *life*, providing not only energy via the tricarboxylic acid cycle and oxidative phosphorylation, but also many other essential services to the cell. These include lipid metabolism and the ability to live in the toxic world of oxygen.

Rudyard Kipling, in his *Just So Stories*, provided fanciful explanations of biological phenomena, such as how the elephant got its trunk, that were scientifically untestable. In evolutionary biology, too, we have similarly untestable *Just So Stories*, such as the one that follows. But it may have value in helping to frame what is to come, and we will return to this fantasy at later stages in this and other chapters.

About 2 billion years ago, an α-proteobacterium invaded an archeon cell. The subsequent endosymbiotic relationship produced what we now know as a eukaryotic cell, and the bacterium became the first mitochondrion.[1] Although this idea of a grand symbiosis is strongly supported by data, the initial relationship may well have started-off far from cooperatively. After all, the bacteria was *infecting* the archeon.

As we have noted, a very good strategy for host defense against intracellular pathogens is for the infected cell to die. This even seems to apply to single-cell organisms that die altruistically to avoid spreading infection to their identical clone mates.[2] Of course, a pathogen that can prevent such proactive suicide gains the upper hand. The initial infection that led to the formation of the first mitochondrion may therefore have triggered a suicide response in the invaded cell (probably not apoptosis, but a precursor to this form of cell death). In turn, this was checked by the infecting bacterium (Fig. 4.1).

As the bacteria became mitochondria, decisions regarding life and death of the cell may have resided under the control of this evolving organelle. In time, the control moved from mitochondria to the eukaryotic cell as a whole. The mitochondria, however, remained the focus for the effects.

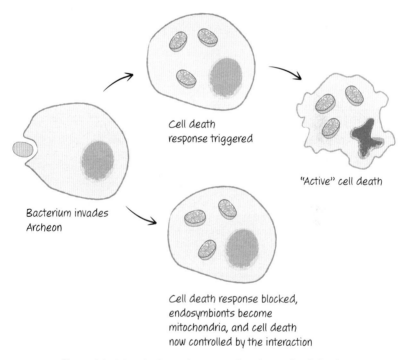

Figure 4.1. A *Just So Story* about mitochondria and cell death.

---

[1] This is the endosymbiont hypothesis of mitochondrial evolution.

[2] There is a literature on active cell death (although not apoptosis) in bacteria and evidence that this is an effective response to infection by phage.

HOW MITOCHONDRIA ACTIVATE CASPASES

The mitochondrial pathway of apoptosis is currently only known to exist in vertebrates, and therefore our initial consideration concerns such animals. As we know, apoptotic pathways involve mechanisms that activate an initiator caspase, which in turn activates the executioner caspases to orchestrate cell death. In the mitochondrial pathway, the initiator caspase is caspase-9.

In Chapter 3, we discussed how activation of initiator caspases by dimerization of inactive monomers requires an adapter protein to bind to their prodomains. In the case of caspase-9, the adapter that activates the caspase is APAF1 (apoptotic protease activating factor 1). APAF1 binds to caspase-9 via its CARD domain, which binds to the CARD domain in caspase-9 (Fig. 4.2).

APAF1, like caspase-9, preexists in the cell as a cytosolic, inactive monomer that cannot bind to or dimerize the caspase. Several events must occur to change this during apoptosis. To understand these, it is useful to look at the different domains of APAF1 (Fig. 4.3).

The CARD region of APAF1, which binds to caspase-9 (via its own CARD), and the oligomerization domain are both presumed to be buried in the inactive APAF1 monomer. The NACHT[3] domain contains a nucleotide-binding site that is inaccessible,

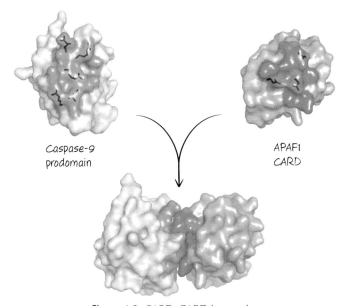

Figure 4.2. CARD–CARD interaction.

---

[3] NACHT is an acronym based on several proteins that contain this domain: NAIP, CIITA, HET-E, and TP1. Although these proteins do not concern us here, APAF1 contains this domain. It is also referred to as a nucleotide-binding domain (NBD).

Figure 4.3. Domains of APAF1.

and we think this is because access to it is blocked by the WD domain.[4] If the WD domain is removed, the nucleotide deoxy-ATP (dATP) can bind to the nucleotide-binding site in the NACHT domain. This triggers a conformational change, exposing the oligomerization and CARD domains. APAF1 then assembles into the complex shown in Figure 4.4. The center of the APAF1 oligomer contains the CARDs that recruit caspase-9 to activate it. This APAF1–caspase-9 complex is called the "apoptosome."

Removal of the WD region results in formation of the apoptosome, but this is not what occurs in cells undergoing apoptosis. Instead, another protein binds to the WD region, producing a conformational change that exposes the nucleotide-binding site (presumably by moving the WD region away from it). The protein responsible is cytochrome c, a protein present in mitochondria where it has a central role in electron transport and energy production.

Cytochrome c is a nuclear-encoded protein that is synthesized in the cytosol as apo-cytochrome c and then transported into mitochondria to the space between the

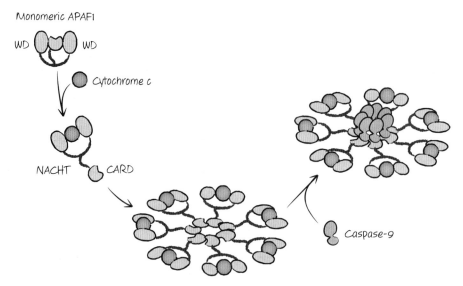

Figure 4.4. Apoptosome formation in vertebrates.

---

[4] WD domains (also called WD40) contain multiple motifs of ~40 amino acids that often end with a tryptophan-aspartic acid. Many proteins have WD domains that have functions in signaling and cell cycle control (and in the case of APAF1, apoptosis).

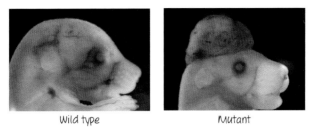

Wild type   Mutant

Figure 4.5. Mice without APAF1, caspase-9, or caspase-3 or with a cytochrome K72A mutation have forebrain overgrowth.

inner and outer mitochondrial membranes (the intermembrane space). Here, the enzyme heme lyase attaches a heme group to create the mature protein (holocytochrome c). Apo-cytochrome c cannot activate APAF1, but holocytochrome c (which can do so) is normally sequestered away from the cytosol (and APAF1) by the mitochondrial outer membrane. For apoptosis to occur by the mitochondrial pathway, this barrier must be disrupted.

The ability of cytochrome c to engage APAF1 to induce apoptosome formation and caspase activation is independent of its function in electron transport. We can introduce mutations in a particular amino acid (lysine 72) that do not affect electron transport but impair apoptosome formation. Mammalian cells lacking cytochrome c do not activate caspases when the mitochondrial pathway is engaged, nor do cells in which cytochrome c is mutated at this key residue.

Mice lacking APAF1 or caspase-9 often die during development or just after birth, owing to extensive developmental mutations. These include large outgrowths of the brain because of excess neurons. Cells from these mice do not activate caspases in response to signals that engage the mitochondrial pathway of apoptosis. These defects in development are also seen in mice lacking caspase-3. In addition, a mouse has been generated in which cytochrome c was mutated at lysine 72. This mutation produced an animal with the same defects as in those lacking APAF1, caspase-9, or caspase-3 (Fig. 4.5). This serves as a formal demonstration that the mitochondrial pathway of apoptosis requires the function of cytochrome c.

DISRUPTING THE MITOCHONDRIAL OUTER MEMBRANE

If cytochrome c is to trigger apoptosis by engaging APAF1, it must move from the mitochondrial intermembrane space to the cytosol, where both APAF1 and caspase-9 reside. It is able to do this during apoptosis because upstream signals that induce cell death cause the outer membranes of all (or nearly all) mitochondria in the cell to become permeable by a process called mitochondrial outer membrane permeabilization (MOMP). This results in the release, by diffusion, of any soluble molecules residing in the intermembrane space, including cytochrome c. An example is shown in Figure 4.6.

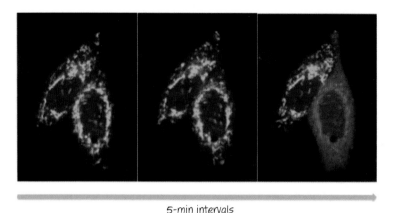

Figure 4.6. MOMP. A cell expressing cytochrome c–green fluorescent protein (GFP) fusion, undergoing MOMP in response to an apoptosis-inducing stress (*left to right*). As MOMP occurs, the fluorescence goes from localization in the mitochondria to diffuse throughout the cytoplasm. The other cell (to the *left*) underwent MOMP at a later time. The time between images is 5 min, with the first image taken several hours after the initial stress.

Through time-lapse imaging of cells expressing fluorescent fusion proteins, we know that during apoptosis, MOMP is sudden, rapid, and irreversible. That is, when a cell is induced to undergo apoptosis, an indeterminate time passes and then suddenly nearly all of the mitochondria undergo MOMP within a very short time period (about 5–10 min). Shortly after this, caspases become active, leading to apoptosis.

Because MOMP does not involve a loss of integrity of the inner mitochondrial membrane, mitochondrial function is not destroyed by this process, although electron transport is greatly reduced because cytochrome c becomes diluted in the cell. However, as we saw in Chapter 2, as executioner caspases become active they gain access to proteins on the inner membrane. Consequently, complex I of the electron transport chain is destroyed and mitochondrial physiology is dramatically altered.

OTHER PROTEINS ARE RELEASED WITH CYTOCHROME C

When MOMP occurs during apoptosis upstream of caspase activation by the apoptosome, all soluble proteins of the intermembrane space are free to diffuse out of the mitochondria and into the cytosol. Of these hundreds of different proteins, some have roles in apoptosis and possibly other forms of cell death.

Among these is a protein called Smac (second mitochondrial activator of caspases; also called Diablo). This protein performs a function in caspase activation that is distinct from that of cytochrome c. Like the latter, it is produced by a nuclear gene, and the protein is imported into the mitochondrial intermembrane space. In the process,

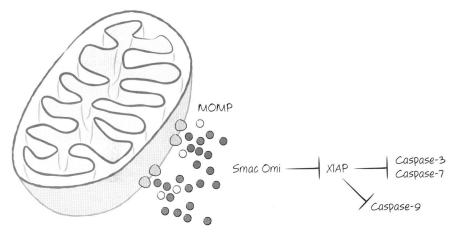

Figure 4.7. Apoptosome–XIAP/caspase-9, Smac.

a signal peptide is proteolytically removed, revealing a short sequence at the new amino terminus with an important function.

Remember that vertebrate cells express XIAP, an endogenous inhibitor of caspase-9 and the executioner caspases. This can block caspase activation and apoptosis even if cytochrome is released to trigger the formation of the APAF1 apoptosome. The function of Smac is to prevent XIAP from exerting this inhibitory effect. The amino terminus of Smac binds to the same region of XIAP that binds to caspase-9, allowing caspase activation to proceed. This is illustrated in Figure 4.7.

Smac is a large protein dimer. Is the only function of this protein to present the small amino-terminal peptide to XIAP following MOMP? It seems unlikely, but no other mitochondrial functions of this protein have been identified. Peptides and drugs that mimic the amino terminus of Smac bind to and inhibit two other IAPs (in addition to XIAP), cIAP1 and cIAP2, that have roles in other types of (nonapoptotic) signaling. However, at this point we simply do not know if Smac has a function beyond that in apoptosis.

Smac is not the only protein with this XIAP-neutralizing activity that is released following MOMP. Another is Omi (also called HtrA2), which like Smac has an amino-terminal sequence that inhibits XIAP. Omi has other functions in the mitochondria that appear to be conserved not only in animals but in other eukaryotes as well.[5] In some animals, Omi does not carry the XIAP-neutralizing sequence at all (one example is the Omi found in cows). We return to Omi and Smac in Chapter 6 when we consider another pathway of vertebrate apoptosis.

---

[5] This is based on studies in organisms lacking the gene. In knockout mice, for example, the animals display cell death in the brain and immune systems ~1 mo after birth, but the reasons for this are not fully elucidated. It is clear, however, that this is not due to the role of Omi discussed here.

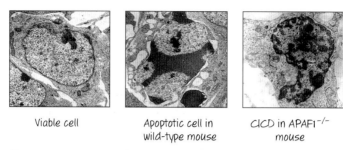

Viable cell   Apoptotic cell in wild-type mouse   CICD in APAF1$^{-/-}$ mouse

Figure 4.8. Apoptosis and CICD in interdigital webs of developing mice.

MOMP CAN CAUSE CASPASE-INDEPENDENT CELL DEATH

Cells in which the mitochondrial pathway of apoptosis is engaged, but in which caspase activation is disrupted or blocked, can still die as a consequence of MOMP. This mode of cell death is often referred to as caspase-independent cell death (CICD). The name is a bit problematic, because *any* cell death that is not apoptotic is generally "caspase independent." But CICD has come to imply that MOMP is involved. In conditions under which MOMP is inhibited, CICD does not occur (we discuss the regulation of MOMP in detail in the next chapter).

Mice lacking APAF1 have a number of developmental abnormalities, as we have seen. However, in many tissues in which developmental apoptosis[6] would normally occur, these mice also display cell death but it has a different appearance. In particular, the cells die without the characteristic appearance of apoptosis. An example is cell death that is seen in the interdigital webs of the developing mice, an event required for the formation of digits (Fig. 4.8).

In fact, when we look more closely at developmental cell death in wild-type animals, some cell deaths more closely resemble CICD than apoptosis. It may be that in some cells, APAF1 or the caspases are not efficiently engaged following MOMP, but because the cell dies anyway, it does not matter. An example of what appears to be CICD in a wild-type tissue is also shown in Figure 4.8.

Not all cells undergo CICD in response to MOMP if caspases are not active. In neurons, for example, cells can survive MOMP if caspases are not activated, and the cells can eventually recover. In culture, most cells that undergo MOMP are generally doomed, but under certain conditions cells can recover if caspase activation is blocked or disrupted. CICD therefore appears to be less efficient than apoptosis, which may help to explain the developmental abnormalities in knockout mice that cannot engage caspase activation via the mitochondrial pathway.

---

[6] Discussed in more detail in Chapter 10.

MECHANISMS OF CICD

One likely explanation for CICD is "mitochondrial catastrophe." Once MOMP occurs, the outer membranes of the mitochondria are compromised, and therefore all of the soluble proteins from the intermembrane space become severely diluted, affecting mitochondrial function. ATP production, lipid biogenesis, and other important functions of mitochondria no longer occur efficiently, and the cell reaches a point of no return and expires.

An alternative explanation for CICD is that some of the proteins released following MOMP can kill the cell regardless of whether caspases are activated. Two of these potential killers are endonuclease G and apoptosis-inducing factor (AIF).

Endonuclease G is a mitochondrial enzyme that can cleave DNA between nucleosomes, similarly to CAD/DFF40 (discussed in Chapter 2). Cells lacking CAD or iCAD, or cells in which caspases are inhibited, generally fail to fragment their DNA during cell death. It is not clear, therefore, that endonuclease G cleaves nuclear chromatin during either apoptosis or CICD, and the role of endonuclease G in CICD is not well established. Nevertheless, it remains possible that limited cutting of DNA by this enzyme contributes to CICD; at this point, we cannot say with certainty that it is involved.

The case for AIF is more intriguing. AIF is an essential protein that appears to be important for the proper assembly and function of the mitochondrial electron transport chain, and homologs of AIF are found throughout the eukaryotes. It resides in the intermembrane space and is tethered to the inner membrane. Following MOMP, proteases gain access to the intermembrane space as we described and then free AIF from its tether. Caspases can do this, as well as other proteases, such as calpain. It has been suggested that AIF then locates to the nucleus to effect CICD, possibly by causing DNA fragmentation (Fig. 4.9).

Lack of AIF appears to be incompatible with tissue development, and cells lacking AIF are severely compromised, which makes it difficult to rigorously test the importance of AIF in CICD. Although there are many studies concluding that AIF release from mitochondria is a cause of cell death in various systems, this remains controversial.

MITOCHONDRIAL PERMEABILITY TRANSITION

How does MOMP occur? If isolated mitochondria are exposed to high concentrations of calcium, this causes a channel—the permeability transition pore (PTP)—to open in the mitochondrial inner membrane. As a result, the transmembrane voltage potential across the inner membrane immediately dissipates, solutes enter the matrix, and water swells the mitochondria until the inner membrane ruptures the outer membrane. This change in the inner membrane is referred to as the mitochondrial permeability transition (MPT).

Many signals in addition to calcium can cause the MPT, including reactive oxygen species, changes in cellular pH, and certain drugs. Because of this, it was widely be-

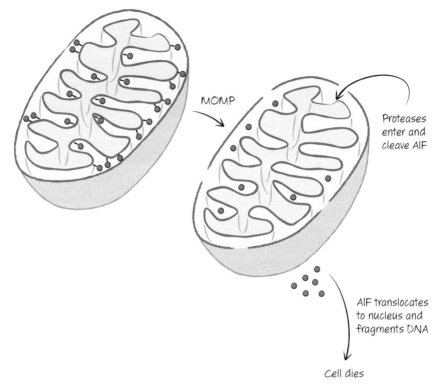

Figure 4.9. AIF model of CICD.

lieved that the MPT was the event that initiates apoptosis by breaking the outer membrane and releasing cytochrome c (and other proteins). This view persists in some quarters. However, although the MPT is likely to have important roles in some forms of cell death, the available evidence is that it has no (or little) role in most forms of apoptosis, including the mitochondrial pathway of apoptosis, as we will see.

Unfortunately, we know very little about what comprises the PTP. Most schemes representing the PTP show several characterized molecules involved in forming a channel through both the inner and outer membranes, such as that depicted in Figure 4.10.

In this view, the PTP is mostly composed of the adenosine nucleotide transporter (ANT, which shuttles ADP and ATP across the inner membrane) with roles for the voltage-dependent anion channel (VDAC) in the outer membrane. The PTP forms when ANT opens a channel in the inner membrane. However, mitochondria from cells lacking different ANTs display normal permeability transitions (as well as apoptosis), as do cells lacking different VDACs. So currently, we do not know which membrane molecules comprise the PTP. One model suggests that many different proteins in the inner membrane might have this activity.

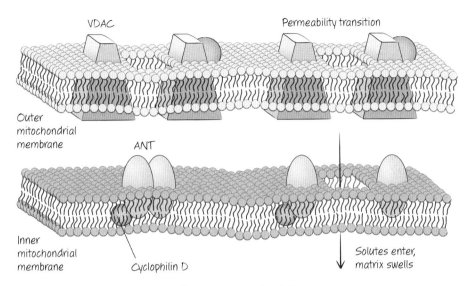

Figure 4.10. Not the PTP.

One generally agreed-upon component of the PTP is a matrix protein called cyclophilin D. Cyclophilin D is a peptidylprolyl isomerase, an enzyme that reconfigures proline residues in proteins.[7] Mitochondria from mice lacking cyclophilin D display a defective MPT response to calcium (although at much higher concentrations of calcium, an MPT can be detected). Significantly, however, no defects in apoptosis, either developmental or induced, can be detected in these mice, and they are developmentally normal. Interestingly, the mice are strikingly resistant to a nonapoptotic form of cell death because of ischemic injury (discussed in more detail in Chapter 8).

Given these observations, MPT is unlikely to be a major mechanism of MOMP in apoptosis. The calcium levels needed to trigger MPT are not achieved in cell signaling but may arise under pathological conditions. A more likely set of molecules that control MOMP in the mitochondrial pathway of apoptosis is the subject of the next chapter.

APOPTOSOMES OF FLIES, WORMS, AND OTHER BEASTS

The adapter proteins responsible for the activation of initiator caspases in the fly (Dronc) and nematode (CED3) are homologs of APAF1 (Fig. 4.11). In *Drosophila*,

---

[7] Cyclophilin D interacts with ANT and other proteins, and it has been suggested that its reconfiguration of prolines in the target proteins is what opens the PTP.

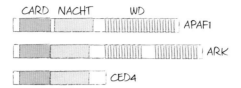

Figure 4.11. APAFs in humans, flies, and worms (domain structures).

this homolog is called APAF1-related killer (ARK) and in *Caenorhabditis elegans*, it is CED4.

In each case, a CARD in the adapter protein binds to the CARD in the prodomain of the caspase to activate the latter. However, unlike APAF1, neither of these appears to be activated by cytochrome c.

For CED4, the reason for this is pretty clear: CED4 does not possess a WD region, and as a consequence, it can spontaneously oligomerize to form an octameric apoptosome and activate CED3. In healthy cells, this is prevented by another protein, CED9, that holds CED4 as an inactive monomer. Curiously, this occurs on the surfaces of mitochondria for reasons that are not known (but we speculate on this in the next chapter). During apoptosis, CED4 is released from CED9. CED4 then oligomerizes, recruits CED3, and activates the caspase to promote apoptosis (Fig. 4.12).

In flies, the APAF1 homolog, ARK, appears to be constitutively active. That is, the protein might spontaneously oligomerize to bind and activate the initiator caspase Dronc. We say "might" because ARK, like APAF1, has a WD region, and therefore, another molecule (a protein?) may have a role in the activation of ARK, but currently, this is not known.[8] However, if ARK is constitutively active, what holds apoptosis in check? The answer is the *Drosophila* IAP, DIAP1, which prevents Dronc activation. It is the disruption of the DIAP1–Dronc interaction that triggers apoptosis. For now, we can create the scheme shown in Figure 4.13 for apoptosis in the fly.

What do the apoptotic pathways in these animals say about our *Just So Story* of an ancient role for mitochondria in controlling cell death? Is the nematode pathway ancestral, giving rise to the arthropod pathway in insects, and then the mitochondrial pathway in vertebrates? If so, then our story is simply a fantasy without value.

Flies did not evolve from nematodes, however, and humans did not evolve from flies. It is possible that the mitochondrial pathway, as it appears in vertebrates, is the ancestral mechanism. For some reason, nematodes and insects may have subsequently lost it.

---

[8] There is some controversy about this. Genetic evidence supports a role for cytochrome c in the activation of caspases in spermatogenesis in the fly and perhaps apoptosis in some cells. However, experiments with *Drosophila* cells and cell extracts do not support a role for cytochrome c or its release from mitochondria in apoptosis.

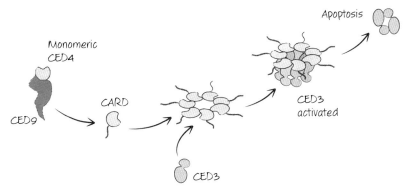

Figure 4.12. Caspase activation in nematodes. Remember that this illustration is misleading: CED4 oligomerizes near its nucleotide-binding region into an octomeric apoptosome; the CED4 CARD binds to the CARD of CED3, which is then activated by proximity.

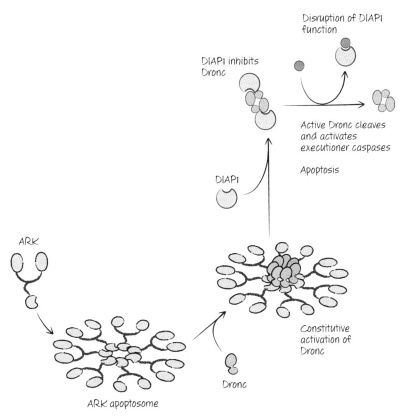

Figure 4.13. Caspase activation in *Drosophila*. ARK is constitutively active, and forms the apoptosome, which activates Dronc. Dronc is held inactive by DIAP1. Disruption of DIAP1 permits the active caspase to cleave executioner caspases, and apoptosis proceeds.

Currently, to guide us we have only sequence data from different animal phyla. APAF1 homologs are found throughout the animal kingdom, and in those examined so far, only in nematodes do any of these lack the WD region that interacts with cytochrome c in the vertebrate pathway. At this point, we do not know if any of these actually interact with cytochrome c, and we therefore cannot draw too much by way of conclusions. However, what we *can* say is that the pathway in nematodes is unlikely to be ancestral.

IAP INHIBITION IN *DROSOPHILA* APOPTOSIS

As we have discussed, the APAF1 homolog in flies appears to be constitutively active, and apoptosis is controlled by the action of DIAP1 to block the initiator caspase Dronc.

In flies, MOMP does not occur upstream of caspase activation, and therefore no IAP antagonist is released from mitochondria to promote cell death. Instead, proteins that use a strategy similar to that of Smac and Omi are transcriptionally expressed and neutralize DIAP1 to cause apoptosis. These proteins are encoded within a complex of functionally related genes that all have suitably morbid names including Grim, Reaper, Sickle, and the less evocative Hid (because it was originally identified in a different setting). These proteins, like Smac (and other IAP antagonists), share the small amino-terminal region that allows them to bind to the IAP and take it out of action. The resulting *Drosophila* pathway for apoptosis is shown in Figure 4.14.

Smac, Omi, and these *Drosophila* proteins all interact with IAP proteins in similar ways, through the binding of their amino-terminal sequences; Figure 4.15 shows these

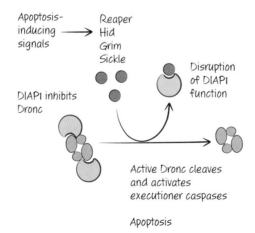

Figure 4.14. Apoptotic signals induce the expression of inhibitors of DIAP1.

Smac	AVPI
Omi	AVPS
Reaper	AVPI
Hid	AVPF
Grim	AIAY
Sickle	AIPF
Caspase-9	ATPF

Figure 4.15. IAP-binding sequences in humans and *Drosophila*. The amino-terminal sequences of Smac and Omi are produced following cleavage of the mitochondrial localizing sequence in mitochondria. That of caspase-9 is produced by caspase cleavage.

sequences for comparison. The same sequence is found in caspase-9, and the binding of this sequence to XIAP is necessary for inhibition of the caspase. The ability of pro-apoptotic proteins to neutralize IAP proteins therefore appears to be simple competition with the caspases for binding. If so, this small peptide region should be sufficient to produce the effect. This appears to be true in experimental systems.

So far, we have considered only the part of the mitochondrial pathway of apoptosis that is downstream from MOMP, and MOMP is clearly an important event that determines life or death, at least in vertebrate cells. What causes MOMP and how it is regulated are considered next.

CHAPTER 5

The Mitochondrial Pathway of Apoptosis, Part II
The BCL-2 Protein Family

CONTROLLING MOMP

We are now near the top of the mitochondrial pathway of apoptosis, at the point before the mitochondrial outer membrane permeabilizes to release proteins such as cytochrome c into the cytosol. This is where the major decisions are made that determine whether a cell will die by engaging this pathway. And these decisions depend on interactions among members of the BCL-2 family of proteins.

The BCL-2 proteins, named for the first family member to be described (B cell lymphoma-2), are a collection of related molecules found throughout the animal kingdom. They share only limited sequence similarity, except in short regions called BCL-2 homology (BH) domains, and can be grouped according to which of these domains they carry and, as we will see, by their functions. A list of several BCL-2 proteins in mammals and their BH domains is shown in Figure 5.1.

There are three "flavors" of BCL-2 proteins. The *pro-apoptotic BCL-2 effectors* promote apoptosis by causing MOMP (mitochondrial outer membrane permeabilization). These are the proteins that essentially make the holes in the outer mitochondrial membrane. The *anti-apoptotic BCL-2 proteins*, which include BCL-2 itself, prevent apoptosis by preventing MOMP. The third group is a subfamily of proteins that promote apoptosis by regulating the other two types of BCL-2 molecules. These are the *BH3-only proteins*, which share only the BH3 domain—hence their somewhat unfortunate designation.

The BCL-2 proteins are an alphabet soup of names, some sounding nearly identical and others bordering on the unpronounceable (BMF is "bimf," if that helps). The acronyms have ceased to have any real meaning, and we will treat them here as sim-

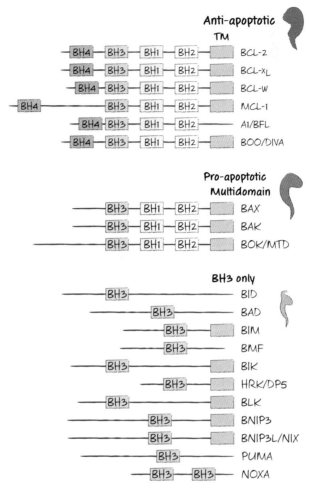

Figure 5.1. BH domains and the BCL-2 family. This is an incomplete list. (TM) Transmembrane domain.

ple (if confusing) names.[1] Nevertheless, these molecules are essential to the mitochondrial pathway of apoptosis.

BAX AND BAK ARE THE EFFECTORS OF MOMP

At first glance, BAX and BAK seem to be quite different proteins, sharing little sequence similarity outside the short BH1, BH2, and BH3 domains. BAX is generally soluble in

---

[1] To make matters a bit worse, many of the BH3-only proteins have distinct gene names (the gene encoding PUMA is *bbc3*).

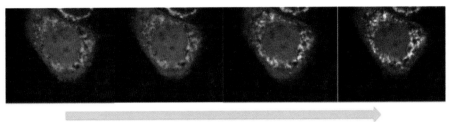

10-min intervals

Figure 5.2. BAX moves from the cytosol to the mitochondrial outer membrane during apoptosis. A cell expressing BAX-GFP (green fluorescent protein) and a mitochondrial-localized RFP (red fluorescent protein) is induced to undergo apoptosis. BAX (green) moves onto the mitochondria (red). The first image was taken several hours after the initial stress.

the cytosol but moves to mitochondria when apoptosis is induced (Fig. 5.2). In contrast, BAK is tethered to mitochondria in the cell by its carboxy-terminal region.

When BAX and BAK are activated, they bury themselves in the mitochondrial outer membrane, where they form oligomers of multiple sizes. Studies using artificial membranes have shown that these oligomers can form holes in membranes (technically, these are not pores or channels, both of which are much more organized). These openings are capable of allowing large molecules to pass through the membrane.

Mice engineered to lack BAX or BAK develop normally, although BAX-deficient mice have reproductive problems, and cells from these animals generally undergo MOMP and apoptosis normally. However, animals lacking both BAX and BAK are another story. These have severe developmental defects that are usually lethal to the embryo. Cells from such double-deficient animals do not undergo MOMP or engage the mitochondrial pathway of apoptosis. Therefore, this seems to be a striking case of molecular redundancy—either BAX or BAK can effect MOMP, but at least one must be present.

How do BAX and BAK perturb the outer mitochondrial membrane to cause MOMP? Structurally, the proteins seem similar, but we do not know what structures they attain once they have been activated and insert themselves into the outer membrane. Currently, it is thought that BAX or BAK oligomers may disrupt the organization of the lipids in the membrane to form holes.

Another BCL-2 family protein, BOK, is often listed among the pro-apoptotic effectors (together with BAX and BAK). BOK can kill cells when it is artificially overexpressed, but there is no current evidence that suggests that it can act like BAX or BAK to effect MOMP. The function of BOK remains a mystery. Another mystery is that of the three proteins BAX, BAK, and BOK, only BOK is often deleted in cancers. As we will see in Chapter 11, apoptosis is activated by oncogenic transformation and suppresses cancer unless the apoptotic pathway is disrupted.

Anti-apoptotic BCL-2 Proteins Prevent MOMP

The anti-apoptotic proteins BCL-2, BCL-xL, MCL-1, and A1 (among others) all act to prevent MOMP and block the mitochondrial pathway of apoptosis. These may work simply by preventing the oligomerization of BAX and BAK, which indeed they do. But *how* they do this brings us to the edge of waters made murky by controversy.

At first, everything seems pretty clear. The anti-apoptotic BCL-2 proteins bind to active BAX and BAK. More specifically, they bind to the BH3 regions of BAX and BAK. The structure of BCL-xL bound to a peptide corresponding to the BH3 of BAK has been solved and is fairly informative (Fig. 5.3).

The BH3 domain of BAK binds to a groove in BCL-xL formed from α-helices containing its BH1, BH2, and BH3 domains (the so-called BH groove[2]). This binding pocket is present in all of the anti-apoptotic BCL-2 proteins and is clearly critical for their anti-apoptotic function, because mutations affecting it can destroy the ability to block MOMP and apoptosis.

Here is the first bit of murk. The BH3 regions of BAX and BAK are not exposed in the native proteins (Fig. 5.4), so how could anti-apoptotic BCL-2 proteins bind to them? Indeed, if anti-apoptotic BCL-2 proteins are mixed with BAX or BAK, they do not bind

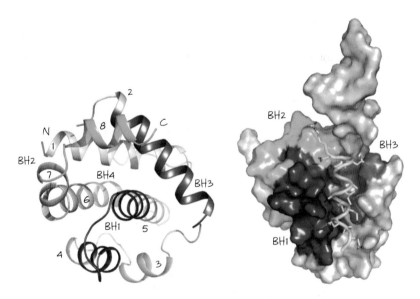

Figure 5.3. Structure of BCL-xL, a globular protein with eight α-helices (numbered 1–8, *left*). The BH1–4 regions are also shown. The structure at *right* shows the groove formed by BH1, BH2, and BH3, bound to a BH3 peptide from BAK (light blue).

---

[2] This is also called a "BC groove," with the "B" referring to the binding of BH3 regions of target proteins and the "C" referring to the binding of the protein's own carboxyl terminus in some of the structured BCL-2 proteins. Here, we use "BH groove" because it is formed from the BH1, BH2, and BH3 domains.

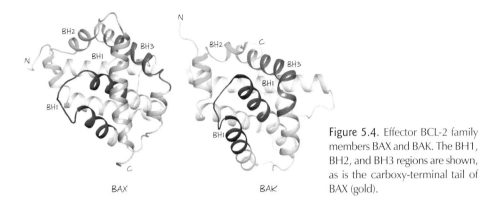

Figure 5.4. Effector BCL-2 family members BAX and BAK. The BH1, BH2, and BH3 regions are shown, as is the carboxy-terminal tail of BAX (gold).

them. However, if detergents are added, they now bind readily. This is a useful hint to what is going on: Anti-apoptotic BCL-2 proteins and pro-apoptotic effectors may only interact when they are embedded in a hydrophobic environment, such as the outer mitochondrial membrane.

So far, so good. Biochemical evidence indicates that as BAK is activated, it exposes its BH3 domain, which opens a groove into which the BH3 domain of another BAK molecule can bind, forming a dimer. BCL-xL, then, may prevent this step in oligomerization by binding to this exposed BH3 domain and thereby preventing BAK–BAK interactions required for MOMP (Fig. 5.5). By extension, the same may be true for BAX.

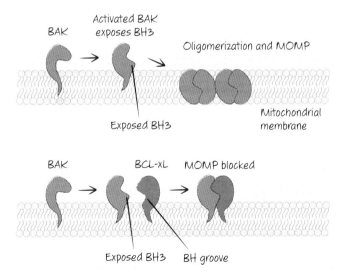

Figure 5.5. BAK activation and inhibition. When BAK is activated, it exposes its BH3 region and appears to create a BH groove. This allows BAK–BAK oligomerization (*top*) or binding of anti-apoptotic BCL-2 proteins to block oligomerization (*bottom*).

Figure 5.6. BAX/BAK specificity of anti-apoptotic BCL-2 proteins. The binding differences are relative, not absolute.

The binding of anti-apoptotic BCL-2 proteins to BAX and BAK is not universal; there is some specificity to the interactions. The BH3 domain of BAK binds to the BH pockets of BCL-xL and MCL-1 very well, but much less effectively to BCL-2. Conversely, the BH3 domain of BAX binds very well to BCL-xL and BCL-2, but poorly to MCL-1 (Fig. 5.6).

But here is the next murky aspect. MCL-1, despite binding poorly to BAX, can block apoptosis very well when apoptosis is mediated by BAX (e.g., when there is no BAK). Similarly, BCL-2 binds poorly to BAK but prevents apoptosis that is mediated by BAK (e.g., when there is no BAX). We can conclude from this that apoptosis is not simply controlled by the balance of anti-apoptotic BCL-2 proteins and the pro-apoptotic effectors BAX and BAK. The latter idea, once called the *rheostat model* (Fig. 5.7), is not quite right; something is missing.

BH3-ONLY PROTEINS PROMOTE MOMP AND APOPTOSIS

The "somethings" missing are the BH3-only proteins. While the BH3 domains of BAK and BAX bind to the anti-apoptotic BCL-2 proteins, the BH3-only proteins also bind to the BH grooves of the anti-apoptotic molecules. The BH3 domain is not well conserved, however, and does not have sufficient sequence characteristics to permit its simple identification by straightforward bioinformatic approaches. Consequently, BH3-only proteins are generally identified by their functions (i.e., binding to anti-apoptotic BCL-2 proteins via a BH3-like region). Therefore, we do not know whether there are more BH3-only proteins waiting to be elucidated or if other unrelated sequences can have similar functions (we return to this idea later). Some examples of bona fide BH3 sequences in BH3-only proteins are shown in Figure 5.8.

Unlike the pro-apoptotic effectors BAX and BAK, most of the BH3-only proteins are intrinsically unstructured, and their BH3 regions may be readily available for binding once they are synthesized, rather than requiring activation by another protein (an

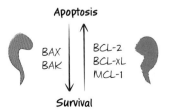

Figure 5.7. The simple rheostat model. Increase in BAX or BAK causes apoptosis, whereas increase in BCL-2, BCL-xL, or MCL-1 results in survival. Problems with this simple model include the specificities of anti-apoptotic proteins for binding to the pro-apoptotic effectors.

```
NIARHLAQVGDSMDRSIP    BID
ECATQLRRFGDKLNFRQK    BIM
RYGRELRRMSDEFVDSFK    BAD
QIARKLQCIADQFHRLHT    BMF
LRAARLKAIGDELHQRTM    HRK
ALALRLACIGDEMDVSLR    BIK
ECATQLRRFGDKLNFRQK    NOXA
EIGAQLRRMADDLNAQEE    PUMA

KLSECLKRIGDELDSNME    BAX
QVGRQLAIIGDDINRRYD    BAK
EVCTVLLRLGDELEQIRP    BOK
     *        *
```

Figure 5.8. BH3 regions of several human BH3-only proteins. BH3 regions of BAX, BAK, and BOK are shown for comparison. Note the conserved leucine and aspartate residues (*) but otherwise little overall consensus.

exception to this is the BH3-only protein BID, discussed in more detail below). These BH3-only proteins can therefore interfere with the ability of an anti-apoptotic protein to bind to the BH3 region of other proteins and thus neutralize them. This depends, of course, on how well the BH3-only protein binds to a particular anti-apoptotic BCL-2 protein (and how much of the BH3-only protein is available). Figure 5.9 shows the relative specificities of several BH3 peptides from BH3-only proteins for different anti-apoptotic BCL-2 proteins.

In a cell expressing both BCL-2 and MCL-1, neutralization of only BCL-2 (e.g., by BAD) might not be sufficient to promote apoptosis unless MCL-1 is also blocked (e.g., by NOXA). This tells us that there may be cooperation between the BH3-only proteins in controlling apoptosis by the mitochondrial pathway. As we will see, however, inhibiting the anti-apoptotic BCL-2 proteins may not be enough to induce MOMP and apoptosis.

SOME BH3-ONLY PROTEINS ACTIVATE BAX AND BAK

As we saw above, the pro-apoptotic BCL-2 effectors BAX and BAK undergo conformational changes, insert into membranes, and oligomerize when they are activated. This results in MOMP unless anti-apoptotic BCL-2 proteins block them. So, if BH3-only proteins are available to block the anti-apoptotic BCL-2 proteins, what activates the effectors? It turns out that some BH3-only proteins have this function too.

BH3 peptides from the BH3-only proteins BID and BIM can trigger BAX or BAK oligomerization and permeabilization of synthetic membranes or isolated mitochondria. In addition, active forms of BID and BIM proteins can activate BAX and BAK to cause MOMP. For this reason, they are referred to as *direct activators*.

Figure 5.9. Specificities of some BH3-only proteins for binding to anti-apoptotic BCL-2 proteins.

Interestingly, mice lacking both BIM and BID do not have the developmental defects seen in BAX–BAK double-knockout mice, and cells from BIM–BID double-knockout mice can undergo MOMP and engage the mitochondrial pathway of apoptosis. Therefore, it is clear that there are other ways to activate BAX and BAK in addition to BID and BIM. Some evidence indicates that the BH3-only protein PUMA is another direct activator of BAX and BAK. This may be true of other BH3-only proteins as well, but we have yet to show this definitively. Most of the BH3-only proteins, however, clearly do *not* have this ability. There might also be other proteins or other types of molecular interactions that activate BAX and BAK. One of these non-BCL-2–family direct activators is considered in Chapter 10.

ACTIVATION OF BAX AND BAK IS BY A "HIT AND RUN" MECHANISM

When BAX or BAK is activated by BID or BIM, the BH3-only direct activators do not remain associated with the pro-apoptotic effector proteins. A physical and transient interaction between BAX and active BID protein has been demonstrated using Förster resonance energy transer (FRET), a method that can detect interactions between proteins by monitoring the energy transfer between fluorescent tags attached to them. These studies showed that BID and BAX interact only when they are in membranes and that this is rapidly followed by the interaction of BAX with other BAX molecules (i.e., oligomerization). This is then followed by permeabilization of the membranes, presumably by BAX oligomers (Fig. 5.10).

Some insights into how the interaction of a direct activator with BAX results in its activation have come from structural studies. A BH3 peptide from the direct activator BIM chemically modified to force it into a stable α-helix binds to BAX in solution (Fig. 5.11). Remarkably, the BIM BH3 domain does not bind to BAX in the region corresponding to the BH groove of an anti-apoptotic protein. Instead, it binds to the "back"

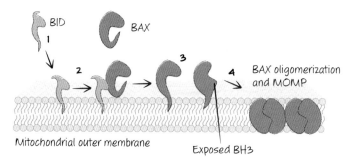

Figure 5.10. Order of events in BAX activation and MOMP. (1) Active BID binds very rapidly to the mitochondrial outer membrane, (2) BID binds to BAX, (3) BAX inserts into the membrane, and (4) BAX binds to BAX molecules, inducing MOMP.

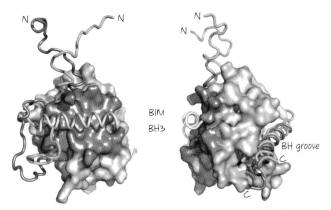

Figure 5.11. The BIM BH3 region binds to BAX and induces conformational changes. Two views are shown. An "α-helix stapled" BIM BH3 peptide (yellow) binds to the "back" face of BAX (brown), opposite the BH groove, where it displaces a loop to cause several subtle rearrangements in BAX. Shown are regions of the free (green) and BIM-bound (blue) BAX that undergo conformational changes.

of BAX. This causes conformational changes in the BAX protein that might serve to expose the BH3 domain of BAX.

Putting this together with the biochemical studies on BAK oligomerization mentioned above (Fig. 5.5), we can envision a process such as that in Figure 5.12. BAX or BAK is activated by interaction with a direct activator protein, such as BID or BIM, causing it to expose its BH3 and a BH-binding groove. As a result, the activated BAX or BAK dimerizes.

This might explain how dimers of BAX or BAK form. For oligomers, we must then envision more complex interactions among the pro-apoptotic effectors, so that they dimerize not only "nose to nose" but also "back to back." Figure 5.5 shows an illustration of such a scheme, but at this point, there is no structural information available to tell us if this represents what really goes on.

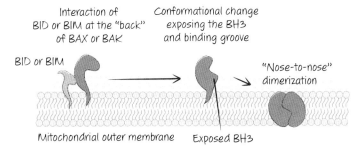

Figure 5.12. Process by which BAX or BAK dimers may form following activation by BH3-only proteins. We do not know how higher-order oligomers form.

TWO MODELS OF BH3-ONLY PROTEIN FUNCTION IN APOPTOSIS

Currently, there are at least two different models for how BH3-only proteins cause MOMP and apoptosis. As we will see, these models are fairly similar on reflection, but they might have profoundly different consequences regarding our ability to manipulate apoptosis for therapeutic benefit.

In the first model, apoptosis is triggered when BH3-only proteins disrupt the interactions between anti-apoptotic BCL-2 proteins and pro-apoptotic effectors BAX and BAK. This model is based on the idea that activation of BAX and/or BAK is neither rate limiting nor a decision point in cell death or survival. When active BAX or BAK is displaced from the anti-apoptotic proteins by the appropriate BH3-only proteins, BAX and/or BAK self-associate and promote MOMP and apoptosis. We refer to this as the *neutralization model* (Fig. 5.13).

The neutralization model is essentially an update of the rheostat model discussed earlier, because the relative levels of functional pro-apoptotic and anti-apoptotic BCL-2 proteins determine if and when apoptosis occurs. In it, the BH3-only proteins drive apoptosis by reducing anti-apoptotic activity. Because BH3-only proteins bind to anti-apoptotic proteins with differing efficiency, the relationships can be complex, but ultimately, life and death in this model are determined by the net anti-apoptotic activity in the cell.

In the second model, activation of BAX and BAK is a critical component of the process that helps to make the decision to undergo MOMP. Here, the anti-apoptotic BCL-2 proteins sequester direct activators of BAX and BAK if such activators are induced. Other BH3-only proteins drive apoptosis by displacing the direct activators, freeing them to trigger BAX and BAK. The latter BH3-only proteins, which lack direct activator function, act as *sensitizers* or *derepressors* in this model. This scheme is shown in Figure 5.14.

In the direct activator/derepressor model, inhibition of anti-apoptotic function will not necessarily cause MOMP and apoptosis unless molecules with direct activator function are present. The major problem with this model is that other than BID and BIM (and a few others), we do not know what activates BAX and BAK.

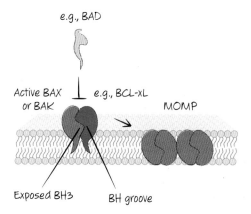

Figure 5.13. Neutralization model for BH3-only function. BH3-only proteins bind to anti-apoptotic BCL-2 proteins to prevent or disrupt the binding of the latter to active BAX or BAK, and MOMP ensues.

THE MITOCHONDRIAL PATHWAY OF APOPTOSIS

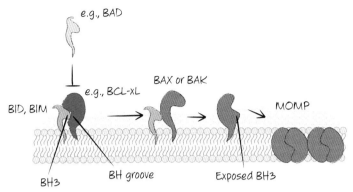

Figure 5.14. Direct activator/derepressor model of BH3-only function. Derepressor BH3-only proteins bind to anti-apoptotic BCL-2 proteins, preventing or disrupting the binding of the latter to direct activators of BAX and BAK. BAX and BAK are then activated, and MOMP ensues.

Although the distinctions between these models are important, the two really are rather similar. BH3-only proteins either displace direct activators of BAX and BAK from anti-apoptotic BCL-2 proteins or they displace active BAX or BAK from such proteins. Ultimately, it is likely that both models will be correct, reflecting what occurs under different circumstances.

BH3-ONLY PROTEINS ACT AS "STRESS SENSORS"

We have seen that BH3-only proteins can promote apoptosis (potentially in two ways), but when do they do this? The different BH3-only proteins have varied tissue distributions, are expressed under different conditions, and are regulated in different ways. They are targets of signal transduction pathways and, therefore, we can think of them as sensors that connect the environment to the mitochondrial pathway of apoptosis.

The following are a few examples of BH3-only proteins functioning as sensors. In the chapters that follow, we return to specific BH3-only proteins in the context of different physiological or pathological situations.

BID IS A PROTEASE SENSOR

Unlike the other BH3-only proteins that have been examined, BID is structured, and it looks similar to anti-apoptotic BCL-2 proteins and pro-apoptotic effectors (Fig. 5.15). The BH3 domain of BID in its native state is unavailable for interaction with other BCL-2 family proteins.

BID has a large flexible loop that can be cleaved by a variety of proteases, including lysosomal proteases (cathepsins), the calcium-activated protease calpain, granzyme B, and caspases (Fig. 5.16). If BID is cut in this linker, the protein can now insert into mitochondrial membranes, and the BH3 domain presumably becomes exposed. BID can

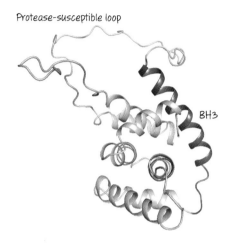

Figure 5.15. Structure of BID, a structured BH3-only protein with one of the most divergent BCL-2 cores. The BH3 region (red) is next to a protease-susceptible loop whose digestion is required for activation.

now interact with anti-apoptotic BCL-2 proteins that sequester it, or it can function to activate BAX and BAK, as we have seen. Therefore, if activating proteases appear in the cytoplasm of a cell, BID cleavage can engage the mitochondrial pathway of apoptosis.

BIM AND BAD ARE SENSORS FOR GROWTH FACTOR SIGNALING

When cells are deprived of growth factors, they often undergo apoptosis. In many cases, growth factor receptor signaling activates a kinase, AKT, that phosphorylates (among other things) the transcription factor FOXO3a. The phosphorylated FOXO3a is sequestered in the cytosol by one of the 14-3-3 proteins that regulate the availability of signaling proteins in the cell, preventing FOXO3a from going to the nucleus. When growth factor signaling is disrupted, such as when growth factors become limiting, FOXO3a is released and triggers the transcription of BIM (Fig. 5.17). Lymphocytes from mice that lack BIM resist apoptosis caused by growth factor deprivation.

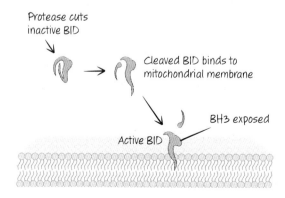

Figure 5.16. BID cleavage and activation.

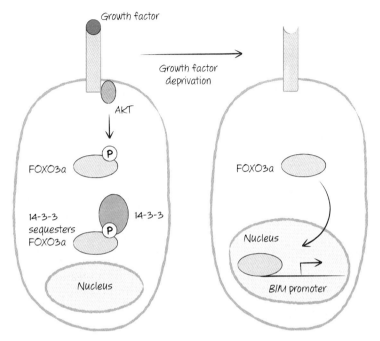

Figure 5.17. Growth factor signaling controls BIM expression.

This is not the only way in which BIM is regulated; the MAP kinase ERK is activated by a variety of signaling mechanisms (including growth factor receptor signaling) and this phosphorylates BIM, which targets it for degradation, thereby promoting cell survival in some settings. Another MAP kinase, called JNK (c-Jun N-terminal kinase), has the opposite effect; when JNK phosphorylates BIM, it prevents its degradation, and therefore JNK can promote apoptosis by BIM. The interplay of the MAP kinases, therefore, can dictate how extracellular signals cause apoptosis through regulating BIM.

The BH3-only protein BAD is directly phosphorylated by AKT, and the phospho-BAD is then sequestered by 14-3-3 proteins (as we saw for FOXO3a). Following growth factor withdrawal, BAD is released and can now neutralize BCL-2 and BCL-xL, releasing active BIM (Fig. 5.18).

BIM and BAD are not the only sensors of growth factor deprivation or related forms of cell stress. As we will see, there are other mechanisms whereby anti-apoptotic proteins, MCL-1 in particular, are regulated under growth factor deprivation, and these, too, contribute to the decision to die.

BIM AND BMF MEDIATE ANOIKIS

Severe disruption of the cytoskeleton induces apoptosis in many cells, and BH3-only proteins are involved in sensing such disturbance. This can happen when adherent

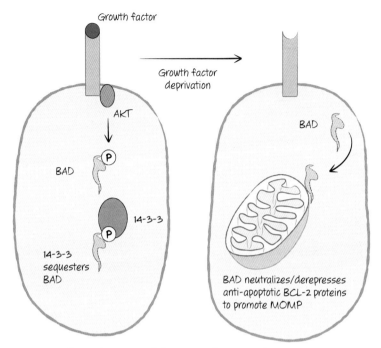

Figure 5.18. Growth factor signaling controls BAD function.

cells lose their attachment to a surface or to basement membranes. The resulting apoptosis is called *anoikis* ("homelessness") (Fig. 5.19).

Dynein motor complexes move along the cytoskeleton in cells. Two BH3-only proteins, BIM and BMF, associate with specific motor complexes, binding to different dynein light chains. BIM appears to associate with microtubules through this interaction, whereas BMF associates with actin filaments. During anoikis, or in response to pharmacological agents that disrupt the cytoskeleton, BIM and BMF are released to promote MOMP and apoptosis.

The control of BIM and BMF by the cytoskeleton is not universal, and in some cells, these BH3-only proteins are not associated in this way. Further, it is not established that the cytoskeletal control of BIM and BMF is the only way in which anoikis occurs.[3] Nevertheless, the interactions with dynein light chains illustrate one way in which BH3-only proteins act as potential sensors.

---

[3] There is at least one other possibility. When adherent cells lose contact with their substrate, they often stick together (this also applies to daughters of a dividing adherent cell lacking substrate attachment). As a consequence, one cell often engulfs the other, a process called *entosis*. The engulfed cell dies. This may be why many adherent cells do not grow in semisolid medium; whenever they divide, one cell is killed by entosis.

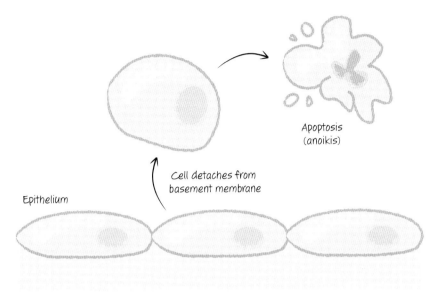

Figure 5.19. Anoikis.

The above are only a few ways in which BH3-only proteins sense apoptotic signals and transduce them to the mitochondrial pathway. Additional examples are discussed in Chapters 10 and 11.

ANTI-APOPTOTIC BCL-2 PROTEINS ARE ALSO CONTROLLED BY SIGNALING

The anti-apoptotic protein MCL-1 is rapidly turned over in cells, and its levels appear to be actively regulated by phosphorylation, ubiquitylation, and degradation. One kinase that phosphorylates MCL-1 to target it for degradation is GSK3, which itself is inhibited by AKT. Therefore, when AKT activity is reduced upon growth factor withdrawal, MCL-1 levels decline as a consequence of GSK3 function. This is illustrated in Figure 5.20. There are almost certainly other kinases that similarly impact the stability of MCL-1.

Other proteins are involved in control of the stability of MCL-1. MULE is one of the E3-ubiquitin ligases that ubiquitylate MCL-1, targeting it for degradation. Intriguingly, MULE has a BH3 domain that binds to MCL-1, and this may interfere with its enzymatic activity. When it is displaced from MCL-1 by a BH3-only protein such as NOXA, MULE can then cause the degradation of MCL-1 by ubiquitylating it.

BCL-2 and BCL-xL are also phosphorylated, but here, the role of phosphorylation is less clear. Phosphorylation of BCL-2 on one particular serine (serine 70) has been reported to increase its activity, but phosphorylation at other sites effectively inhibits it.

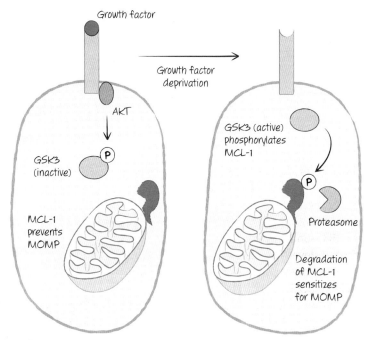

Figure 5.20. Growth factor receptor signaling regulates MCL-1 stability.

Another interesting modification of BCL-xL has been described, involving conversion of two asparagine residues into aspartate and iso-aspartate in the loop between the BH4 and BH1 regions. This appears to affect the stability of the protein and may represent another mode of regulation.

The anti-apoptotic BCL-2 proteins are regulated at the transcriptional level as well, and this may be the major way in which BCL-2 and BCL-xL are controlled. Several transcription factors induce expression of these proteins in different cell types, including nuclear factor-κB (NF-κB), which can produce anti-apoptotic effects in some settings.

BCL-2 PROTEINS IN OTHER ANIMALS

Although the mitochondrial pathway has been unambiguously demonstrated only in vertebrates, BCL-2 proteins are found throughout the animal kingdom. The situation in nematodes is particularly interesting. In *Caenorhabditis elegans*, the APAF1 homolog CED4 is held inactive on the mitochondrial outer membrane by CED9. It turns out that CED9 is a homolog of BCL-2 (Fig. 5.21).

Despite this similarity, mammalian BCL-2 proteins do not sequester APAF1 or affect its activity. Therefore, although BCL-2 and CED9 have similar overt functions (in-

Figure 5.21. CED9 is a BCL-2 protein. The BH1–4 domains of CED9 and BCL-2 are shown for comparison. Structural studies confirm that CED9 and BCL-2 are similar.

hibition of apoptosis), they perform these functions by different biochemical mechanisms.

CED9 has a BH groove like that of the mammalian anti-apoptotic BCL-2 proteins, but this is not involved in binding to CED4. However, another protein, EGL1, binds to this groove and leads to a conformational change in CED9 that causes the release of CED4, promoting apoptosis. This is shown in Figure 5.22.

EGL1 is a BH3-only protein that is expressed in response to transcription factors that are activated developmentally or in response to DNA damage in germ cells in the adult. We can now complete the signaling pathway for apoptosis in nematodes, as shown in Figure 5.23.

As we know, the conformational changes that mammalian BCL-2 proteins undergo during apoptosis require the presence of the outer mitochondrial membrane, and this leads to an intriguing speculation: Perhaps the conformational change that is induced by EGL1 in CED9 to release CED4 is facilitated by the mitochondrial membrane itself. This would help to explain the location of CED9 on mitochondria, despite the absence of MOMP upstream of the activation of the APAF1 homolog CED4 in this animal.

In *Drosophila*, two BCL-2 proteins have been identified on the basis of sequence similarity to mammalian BCL-2 proteins. These are DeBCL (pronounced "debacle") and Buffy (named for the famous vampire slayer), also called DBorg1 and DBorg2, respectively. However, neither has been shown to have any roles in apoptosis in *Drosophila*, which is intriguing, and their functions in flies are unknown. When we consider other roles for the vertebrate BCL-2 proteins later, it may be interesting to remember these.

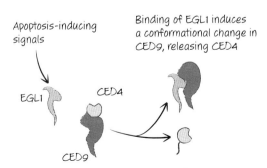

Figure 5.22. EGL1 functions to release CED4.

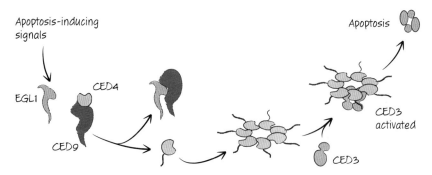

Figure 5.23. Apoptotic signaling in nematodes.

VIRAL BCL-2 PROTEINS

Because viruses have come up with ways to block apoptosis, it may not be surprising that many viruses that infect mammalian cells express anti-apoptotic proteins related to the BCL-2 proteins. Examples can be found in several types of virus, including adenoviruses, herpes viruses, and pox viruses (Fig. 5.24). In each case, the anti-apoptotic protein shares weak sequence similarity to BCL-2, but regions can be identified that correspond to the BH domains.

The structures of a few viral BCL-2 proteins have been solved, and these very closely resemble the cellular anti-apoptotic BCL-2 proteins (Fig. 5.25). In general, however, they lack the regions where regulatory events tend to occur in the cellular proteins, such as phosphorylation; perhaps this is not surprising.

Viruses also express proteins that function like anti-apoptotic BCL-2 proteins (e.g., they bind and neutralize active BAX and BAK) but share no sequence or structural

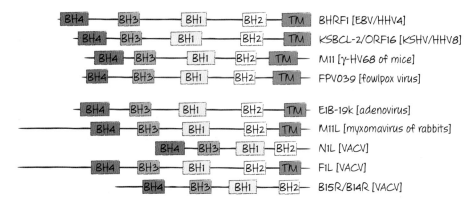

Figure 5.24. Viral BCL-2 proteins. All are from human viruses unless otherwise indicated. Some of the BH domains are based on structural considerations despite low amino acid homology.

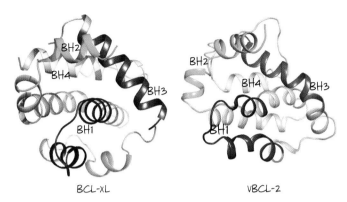

Figure 5.25. Structures of Karposi sarcoma virus BCL-xL and vBCL-2.

similarity with cellular BCL-2. An example is the vMIA protein from human cytomegalovirus. Such proteins can teach us a lot, not only about viral infection, but also how the cellular BCL-2 proteins work.

BACTERIAL TOXINS AND THE RETURN OF THE JUST SO STORY

In the last chapter, we discussed a *Just So Story*, a speculation that apoptosis may have arisen with ancient endosymbiosis that resulted in mitochondria. A piece of that story involves BCL-2 proteins.

The structures of BCL-2 proteins BCL-2, BCL-xL, BCL-w, MCL-1, A1, BAX, BAK, BID, and vBCL-2 are remarkably similar to one another. Intriguingly, a similar structure is also found in some bacterial toxins, including diphtheria toxin β-chain and the colicins (Fig. 5.26).

Each of these proteins undergoes a conformational change to bring a hydrophobic core into a membrane. The structural similarity may simply reflect this common function. But it is intriguing to think about this in the context of our *Just So Story*.

Bacterial toxins of this type are often encoded in a discrete genetic element that actually makes two components: the toxin that creates holes in membranes and an immunity factor that makes the bacterium that produces it resistant to the toxin. When a bacterium harboring this element is stressed, for example, by low levels of nutrients, it produces the toxin. This kills any bacteria that do not have the immunity factor, reducing competition for resources.

The original endosymbiosis, as we speculated, may have begun anything but cooperatively, with the bacterium being a parasite in the infected cell. In this scenario, the inner membrane of what would become the mitochondrion is the bacterial membrane (indeed, the inner membranes of modern mitochondria have lipid compositions

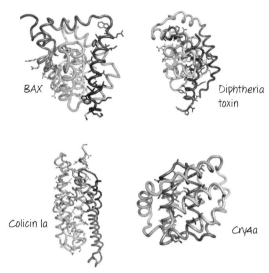

Figure 5.26. Structures of several pore-forming regions of bacterial toxins, compared to BAX. Several α-helical pore-forming toxins have a globular domain reminiscent of the BCL-2 core of BAX. The structures are colored blue to red from the amino to the carboxyl terminus, and the putative pore-forming helices are identified by their side chain sticks.

that more closely resemble bacterial membranes than those of eukaryotes), whereas the outer membrane would have been derived from the infected cell (again, consistent with the composition of modern outer mitochondrial membranes). If stressed, the bacterium might synthesize a toxin that would target the next-nearest membrane: the mitochondrial outer membrane. So, was this the original pro-apoptotic effector?

With time, genetic control of the toxin may have transferred to the nucleus (as did nearly all of the original genes encoding mitochondrial elements, except for those encoding parts of the electron transport chain, ribosomal RNAs, and the tRNAs). Control of the integrity of what became the mitochondrial outer membrane now resided in nuclear genes. And the toxin, perhaps, became the BCL-2 family.

Finally, note that the mitochondrial inner membrane is not permeabilized by active BAX or BAK (we know this, because matrix proteins are not released with the proteins from the intermembrane space). Why not? Is there a distant descendent of the bacterial immunity factor that protects this membrane?

BCL-2–FAMILY PROTEINS HAVE OTHER ROLES IN CELLS

From the perspective of apoptosis, the major role of BCL-2 proteins is to control the integrity of the outer mitochondrial membrane. But these proteins appear to have additional functions in cells, and how these relate to their apoptotic effects and other aspects of cell physiology is emerging as an important area of research. If cell death is the work of the "night crew," then what are the "day jobs" of the BCL-2 proteins?

Given their limited sequence homologies, BH3-only proteins may have numerous functions beyond apoptosis. In at least one case, a day job for a BH3-only protein has been identified. That protein is BAD. In addition to neutralizing some anti-apoptotic BCL-2 proteins, BAD has a role in regulating glucose metabolism in mammals.

We consider here three other day jobs for BCL-2 proteins: control of calcium homeostasis, regulation of mitochondrial dynamics (fission and fusion), and removal of mitochondria during development of some cells. As we will see, these may all have connections to the control of cell death as well as other effects in cells, and of course, there are probably other nonapoptotic roles yet to be identified.

BCL-2 PROTEINS ACT AT THE ENDOPLASMIC RETICULUM TO REGULATE CALCIUM

Inositol-1,4,5-trisphosphate (IP3) is a phospholipid-derived molecule produced in some signaling pathways. IP3 binds to a receptor on the endoplasmic reticulum (the IP3 receptor), causing an efflux of calcium that regulates many different enzymes in the cytosol.

BCL-2 proteins influence the function of the IP3 receptor. Cells that lack BAX and BAK or overexpress BCL-2 exhibit defective calcium efflux in response to IP3. For example, T lymphocytes lacking BAX and BAK do not elevate calcium in response to T cell receptor engagement (which induces IP3 production) and are defective for T cell activation.

How BCL-2 proteins control the IP3 receptor is not entirely clear. BCL-2 binds to the IP3 receptor, but beyond this, the mechanism is obscure. Intriguingly, some BH3-only proteins can induce an increase in intracellular calcium, and it may be that this is via interaction with BCL-2 or another BCL-2 protein. However, we do not know how this contributes to apoptosis or other cellular effects.

MITOCHONDRIAL DYNAMICS ARE INFLUENCED BY BCL-2 PROTEINS

One of the most intriguing day jobs for BCL-2 proteins is in mitochondrial dynamics. Mitochondria do not simply sit around in cells as discrete organelles; instead, they are constantly undergoing active fission and fusion by complex mechanisms that are conserved among the eukaryotes.

The first clue to this day job is what happens during MOMP. At about the same time as the mitochondrial outer membrane becomes permeable and proteins from the intermembrane space are released, mitochondria often appear to undergo extensive fragmentation. At first, it was thought that this fission might contribute directly to MOMP, but that does not seem to be the case, because MOMP can occur in isolated mitochondria without fission.

A second clue is that, in the absence of BAX and BAK, mitochondria are extensively fragmented, and this turns out not to be due to excess fission but rather to a decrease in mitochondrial fusion. In contrast, increasing the expression of anti-apoptotic BCL-xL increases mitochondrial fusion. Therefore, in two different situations where MOMP is prevented (increased BCL-xL and decreased BAX and BAK), mitochondria display either more or less fusion. How can these observations and those of mitochondrial fission upon MOMP be reconciled?

Here is one way. If BAX and BAK promote mitochondrial fusion independently of their role in MOMP, when they become engaged during apoptosis (effectively leaving their day jobs to join the night crew), fusion decreases. Because BCL-xL prevents this (e.g., by sequestering activators of BAX and BAK, letting them remain doing their day jobs), fusion is enhanced.

How all this might occur is not known, but BAX and BAK can associate with a protein called mitofusin-2 that participates in mitochondrial fusion. Remarkably, the C. elegans protein CED9 can also associate with mitofusin-2 in mammalian cells and similarly enhances mitochondrial fusion. CED9 cannot block MOMP in mammalian cells but stimulates mitochondrial fusion under these artificial conditions.

When apoptosis occurs in C. elegans, mitochondrial fragmentation is observed, even though there is no MOMP (as we know, MOMP is not upstream in the apoptotic pathway in these animals). EGL1, which disengages CED9 from CED4, might therefore also take CED9 away from its day job, promoting mitochondrial fusion.

Mitochondrial fragmentation is also observed in *Drosophila* cells undergoing apoptosis. Here, not only is MOMP not upstream in the apoptotic pathway, but the BCL-2 proteins DeBcl and Buffy do not seem to have a role in cell death. Whether they participate in the mitochondrial fragmentation is not known.

Do these changes in mitochondrial dynamics have anything to do with apoptosis? Some studies have indicated that a protein involved in mitochondrial fission, DRP-1, may help to promote MOMP, and cells in which this protein is blocked can become refractory to MOMP and apoptosis. However, this seems to be the case even when mitochondrial fission and fusion cannot occur, and so the role of this protein in MOMP remains obscure.

BCL-2 PROTEINS CAN PARTICIPATE IN THE REMOVAL OF MITOCHONDRIA

Yet another function of the BCL-2 proteins also relates to mitochondria, but in this case, it is the way in which excess or damaged mitochondria are removed. This occurs by autophagy, the "self-eating" process that allows cells to survive nutrient deprivation and remove damaged or unwanted organelles. We have much more to say about autophagy in Chapter 8. Autophagy of the mitochondria (*mitophagy*) involves the creation of a membrane vesicle around the organelle, and fusion of the vesicle with a lysosome results in the digestion of the contents of the vesicle.

In at least some cases, the signals that bring the autophagy machinery to the mitochondria are members of the BCL-2 family, in particular, two closely related BH3-only proteins, BNIP3 and NIX. Neither BNIP3 nor NIX seems to have a significant role

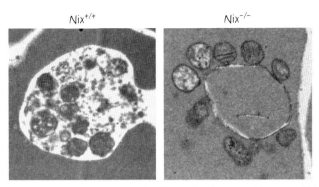

Figure 5.27. Mitochondria in NIX-deficient red cells. In developing wild-type red blood cells, mitochondria are removed by mitophagy (*left*), and these organelles are absent in the mature cells. In NIX-deficient mature red cells, mitochondria persist, many showing signs of damage (*right*).

in apoptosis; instead they appear to function in mitophagy. In mice lacking NIX, mature red blood cells, which normally lack mitochondria because these have been removed, are found to harbor these organelles (Fig. 5.27).

Other studies have suggested that during hypoxia (reduced oxygen levels, a situation in which having smaller numbers of mitochondria is desirable), a transcription factor called HIF-1 is activated that causes expression of BNIP3. The latter then appears to promote the autophagic removal of mitochondria.

How these BH3-only proteins cause mitophagy is not completely clear, but the process may involve binding to BCL-2 and displacing a protein that is bound to it, Beclin-1. Beclin-1 is part of the machinery that initiates autophagy, and its release from BCL-2 on mitochondria may attract components of the autophagy pathway to these organelles (Fig. 5.28). It is particularly noteworthy that Beclin-1 contains a BH3 region and binds to the BH groove in BCL-xL (Fig. 5.29).

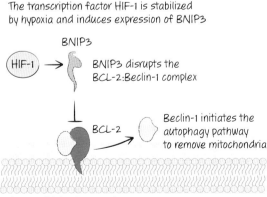

Figure 5.28. Hypoxia leads to removal of mitochondria by autophagy.

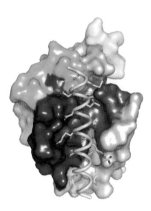

Figure 5.29. Beclin-1 BH3 bound to BCL-xL. The BH regions are colored as in Figure 5.3 and the BH3 peptide is light blue.

We return to the role of Beclin-1 and BCL-2 in autophagy in Chapter 8. For now, it is sufficient to realize that the day jobs of BCL-2 proteins can involve the type of interactions involved in apoptosis (as in this example) or not.

CHAPTER 6

The Death Receptor Pathway of Apoptosis

DEATH RECEPTORS ARE A SUBSET OF THE TNFR FAMILY

So far, we have considered a pathway of apoptosis, the mitochondrial pathway, which is activated in response to cellular stress and some developmental cues. In this chapter, we discuss a different pathway of apoptosis that is triggered by the interaction of extracellular ligands with receptors on the cell surface.

On the surfaces of some mammalian cells (and probably those of other vertebrates as well) are receptors that, when bound by their ligands, engage a pathway of caspase activation and apoptosis that can be distinct from the mitochondrial pathway that we have discussed. These surface molecules are referred to as "death receptors," and their mode of caspase activation is called the death receptor pathway of apoptosis.

The death receptors and the proteins that activate them ("death ligands") are all closely related molecules. The death ligands are members of a larger family of proteins—the tumor necrosis factor (TNF) family—and the receptors for these ligands are members of the TNF receptor (TNFR) family.[1]

The major death receptors are tumor necrosis factor receptor-1 (TNFR1), CD95 (also called Fas and APO-1), DR3 (death receptor 3), the TRAIL receptors (TRAIL receptor-1, also called DR4, and TRAIL receptor-2, also called DR5 in humans; rodents have only one, which resembles DR5), and DR6 (death receptor 6). These are shown in Figure 6.1. The death ligands are also shown and include TNF, CD95-ligand (CD95-L or Fas-L), and TRAIL (TNF-related apoptosis-inducing ligand, also called APO-2L). Although a ligand for DR3 has been identified, this does not readily trigger

---

[1] The TNF and TNFR superfamilies have been given the designations TNFSF and TNFRSF, respectively, followed by a number (those for death ligands and receptors are shown in Fig. 6.1). We have chosen to employ the most widely used names in the text, rather than these accepted designations.

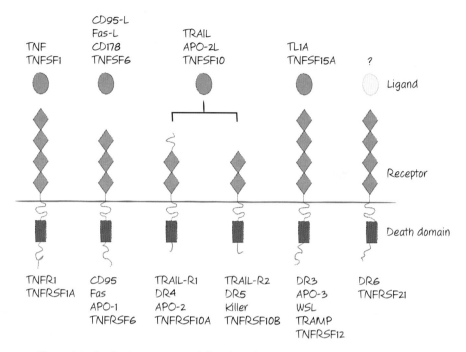

Figure 6.1. The death receptors and their ligands. Several names for each are shown.

apoptosis, and therefore there may be other ways DR3 is engaged or it may not normally function in apoptosis induction. No apoptosis-inducing ligand for DR6 has been described.

All of the TNF-family ligands, including the death ligands, are trimers (an example is shown in Fig. 6.2). This is also true of the receptors.

It is often mistakenly suggested that the unligated death receptors are monomeric and assemble into trimers when bound, but there is very good evidence that the receptors are already trimeric on the cell surface before engaging their ligands. On binding a death ligand, the death receptors expose a region located in the intracellular portion of the receptor. This region contains a "death domain" (DD; see Chapter 3). DDs are found in the death receptors but not in any of the other TNFR-family members. Recall that structurally, they show the death fold also adopted by the DEDs and CARDs found in caspases and their adapters (Chapter 3) (Fig. 6.3). The death domains share no sequence homology with the other death fold domains, but they interact with proteins that share the same domain (DD–DD interactions).

Different death receptors bind to different DD-containing adapter proteins as the next step in the signaling pathway. We begin with the simplest version of this pathway, seen in the case of the death receptor CD95.

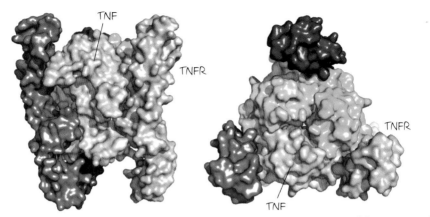

Figure 6.2. Death receptors and death ligands are trimeric. Shown are two views of the structure of TNF (green) bound to the extracellular regions of TNFR.

HOW CD95 TRIGGERS APOPTOSIS

When CD95 is bound by its ligand CD95-L or by some anti-CD95 antibodies that mimic this, the resulting conformational change in the intracellular DD allows it to interact with a small cytosolic protein called FADD. FADD contains a DD through which it binds to CD95 via a DD–DD interaction. The DDs bind to each other at three different interfaces, producing a two-layer structure (Fig. 6.4).

These assemble into arrays as shown in Figure 6.5. This explains why at least two trimers of CD95 must be brought together to produce the complex that leads to apoptosis.

The binding of the CD95 DD to the FADD DD causes FADD to expose a different death fold, a DED. In turn, the exposed DED on FADD now binds to the prodomain

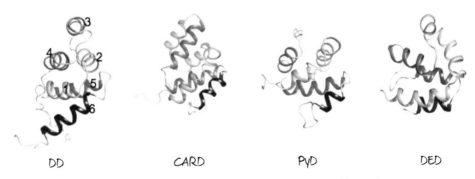

Figure 6.3. Representative structures for distinct death folds are shown.

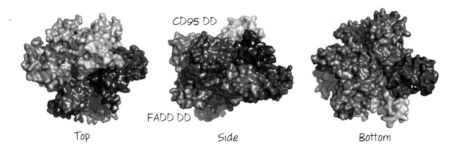

Figure 6.4. The structure of CD95 DD binding to FADD DD.

of the initiator caspase, caspase-8, via one of the two DEDs in this caspase. (It is suspected, but not proven, that the second DED in the initiator caspase helps to hold the first in check to prevent spontaneous interactions.)

In this way, clusters of ligated CD95 cause clustering of FADD, which results in dimerization of caspase-8. This activates the caspase, which then cleaves itself, stabilizing the dimer, as we saw in Chapter 3. Caspase-8 can now cleave and activate the executioner caspase-3 and -7, which in turn orchestrate apoptosis (Fig. 6.6).

The complex composed of ligated CD95, FADD, and caspase-8 is called the CD95 death-inducing signaling complex (DISC). The DISC forms within seconds of CD95 ligation, and apoptosis can proceed within tens of minutes, presumably limited only by the rate of executioner caspase activation and substrate cleavage. However, in some cells, as we will see, this pathway can be slower and somewhat more complicated.

In humans, the DISC can also include caspase-10, which appears to be similarly activated by binding to FADD via DED–DED interactions (note that caspase-10 is not found in rodents). However, at present, there is no compelling evidence that caspase-10 can replace caspase-8 for CD95-induced apoptosis. Human cells lacking caspase-8 are resistant to CD95 ligation (despite the presence of caspase-10).

Not all cells expressing CD95 are sensitive to CD95-induced apoptosis. Another

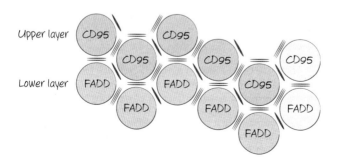

Figure 6.5. Clustering of CD95 and FADD, based on the structure in Figure 6.4. The three different types of interface between different DDs are shown by different lines.

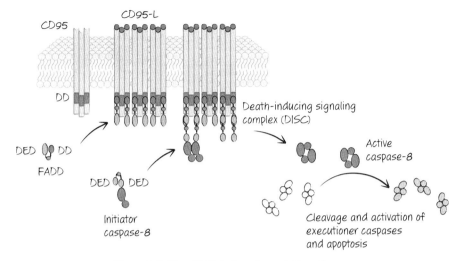

Figure 6.6. The CD95-induced apoptotic pathway.

protein, FLIP, which is related to caspase-8 but lacks an active cysteine, can also bind to FADD in the DISC via a DED–DED interaction (see Chapter 3). If caspase-8 and FLIP are dimerized, caspase-8 becomes activated and cleaves FLIP, but the caspase-8 molecule is not cleaved, and apoptosis does not ensue (Fig. 6.7). The requirement for caspase-8 cleavage for apoptosis to occur is supported by the observation that replacing this caspase with a noncleavable form fails to restore CD95-induced apoptosis, even though the noncleavable caspase-8 dimer is enzymatically active. Cleavage of FLIP may generate a signal with other roles in the cell; alternatively, the caspase-8–FLIP

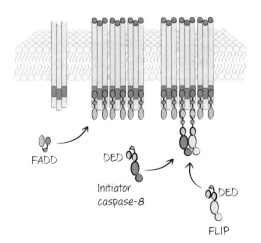

Figure 6.7. FLIP inhibits apoptotic signaling. When FLIP is present, the formation of the DISC recruits it, and this prevents caspase-8 from forming an active homodimer that leads to apoptosis.

dimer may only interact with some substrates but not with others (such as the executioner caspases, which are not activated when FLIP is expressed).

Some herpes viruses express a short form of FLIP (v-FLIP) that blocks CD95-induced apoptosis by binding to FADD and preventing caspase-8 dimerization and activation.

We return to FLIP and its interaction with caspase-8 when we consider another form of cell death in Chapter 8.

DEFECTS IN CD95 SIGNALING CAUSE A LYMPHOACCUMULATIVE DISEASE

Inactivating mutations in CD95, its ligand, or caspase-8 can cause the childhood disease acute lymphoproliferative syndrome (ALPS). This is marked by accumulation of T and B lymphocytes, including a massive increase in an unusual T cell subset not found in normal individuals (for aficionados, this subset is $CD3^+$, $CD4^-$, $CD8^-$, and $B220^+$), presumably because apoptosis that would normally control the numbers of these cells does not occur. Autoimmunity and lymphoid tumors are not uncommon in people with ALPS, and the disease is usually lethal if not treated.

Mice with naturally occurring mutations in CD95 (*lymphoproliferative, lpr*) or CD95-L (*generalized lymphoproliferative disorder, gld*) show the same lymphocyte accumulation seen in ALPS, including accrual of the peculiar T cell type and often become autoimmune (Fig. 6.8). Knocking out CD95 or CD95-L has the same effects, but this is not seen when the deletion of CD95 is restricted to only T cells—it appears that loss of CD95 in another cell type, dendritic cells, is also involved in the disease. Such observations tell us that interactions between CD95 and its ligand are important for homeostasis in the immune system.[2]

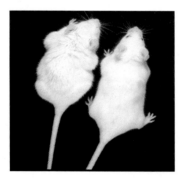

Figure 6.8. *lpr* disease. The mouse on the *left* carries the *lpr* mutation in the CD95 gene, resulting in massive enlargement of the lymphoid organs. The mouse on the *right* is wild type.

---

[2] Intriguingly, background genes influence the susceptibility to the disease, which manifests differently in different mouse strains. This is also true in people. Families that carry dominant-negative forms of CD95 (which can act in a partially dominant manner) can sometimes show ALPS in some children, whereas others are generally normal.

TRAIL AND APOPTOSIS

TRAIL is a TNF-family ligand produced by T cells and other cell types. Like CD95-L, it is a trimer that can trigger apoptosis in target cells bearing the appropriate death receptors. In addition, like its relative, not all cells bearing receptors for TRAIL are sensitive to TRAIL-induced apoptosis, but in this case the mechanisms of resistance are less well understood (although one mechanism probably involves FLIP).

The mechanism of apoptosis induced by ligation of a TRAIL receptor appears similar to that induced by CD95; ligation of the receptor recruits FADD to the death domain on the intracellular region of the TRAIL receptor, and this binds to and dimerizes caspase-8 to activate it. Cells lacking either FADD or caspase-8 are resistant to TRAIL-induced apoptosis.

APOPTOTIC SIGNALING BY THE TNFR1 IS COMPLEX

Two different but related ligands bind to the death receptor TNFR1. These are TNF itself and another TNF-family ligand called lymphotoxin.[3] Ligation of TNFR1 by either causes it to expose its DD and interaction sites for another type of signaling molecule, TNF-receptor associated factor-2 (TRAF-2). The DD of TNFR1 does not bind to FADD, but instead, to another DD-containing adapter protein, TRADD. TRADD helps to stabilize the binding of TRAF-2 and also recruits another molecule to the complex, RIPK1.

RIPK1 is a kinase, but here, it is not the kinase activity that concerns us; the following steps do not depend on its enzymatic function. (We return to RIPK1 kinase activity in another form of cell death in Chapter 8.)

The TRAFs are E3-ubiquitin ligases, and TRAF-2 now functions to polyubiquitinate RIPK1. This involves lysine 63 of the ubiquitin chain and so has a function different from degradation (characteristic of lysine 48 linkages). The lysine 63–linked polyubiquitin on RIPK1 recruits other proteins, including NEMO (also known as I-κB kinase γ, IKKγ). This recruits and activates the IKK complex, which in turn activates the transcription factor NF-κB (Fig. 6.9).

NF-κB induces the transcription of a number of genes involved in cell survival, including FLIP. When NF-κB is activated by TNFR1 ligation, apoptosis does not occur, and instead, the cell responds in other ways—for example, participating in inflammatory responses.

There are additional signaling consequences of this complex. For example, ligation of TNFR1 induces the activation of c-Jun N-terminal kinase (JNK). This phosphorylates the protein Jun, which forms part of another transcription factor, AP-1. TRAIL ligation

---

[3] Both of these also bind to another TNFR-family member that is not a death receptor; TNF binds to TNFR2, and lymphotoxin binds to the lymphotoxin-β receptor. However, because these are not death receptors, we do not further discuss them nor other non-death-receptor members of this extended receptor family.

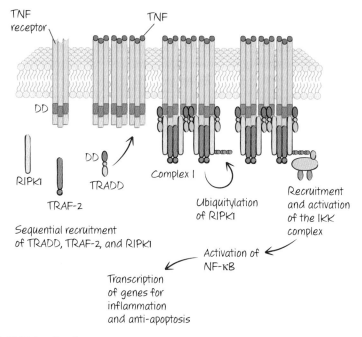

Figure 6.9. TNF signaling for NF-κB, anti-apoptosis, and inflammation. Other proteins and modifications involved in the process are not shown.

can also activate JNK. JNK has other targets in the cell, and its activation can promote apoptosis in some cell types, as discussed below.

If NF-κB is not functional or if survival genes induced by NF-κB are not active, a second signaling step can lead to apoptosis. This happens in some cell types or under conditions where cell signaling is experimentally altered. The modified TRAF-TRADD-RIPK1 complex (called TNFR1 signaling complex I) disengages from the receptor and becomes cytosolic. The DD of TRADD, previously associated with TNFR1, is now exposed, and this recruits FADD. FADD, in turn, binds and activates caspase-8 (forming TNFR1-signaling complex II). The caspase-8 then precipitates apoptosis, as we have seen (Fig. 6.10).

Mice that display defective NF-κB function (because parts of the NF-κB pathway have been knocked out) die during development because of extensive apoptosis in the liver. However, if these animals also lack TNFR1, they survive beyond birth. This tells us that it is apoptotic signaling from TNFR1 that is responsible for liver failure in NF-κB–defective mice.

The rescue of these mice by TNFR1 deficiency tells us that other death ligands, such as TRAIL or CD95-L, do not contribute significantly to the lethal effects of loss of NF-κB signaling during development. Note, however, that activation of NF-κB can

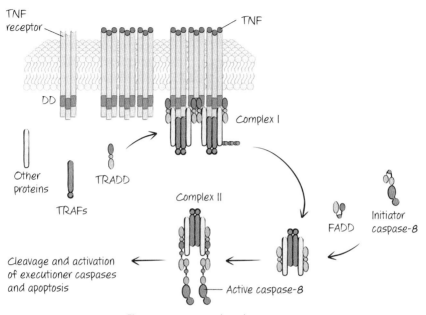

Figure 6.10. TNF-induced apoptosis.

produce resistance to apoptosis induced by ligation of other death receptors by inducing expression of FLIP and other proteins.

There is evidence that other death receptors, including CD95 and the TRAIL receptors, may also form cytosolic signaling complexes such as complex II following ligation, and these resemble complex II in TNFR signaling. At this point, it is not clear that this is necessary for caspase-8 activation and apoptosis in these cases. Furthermore, these and other death receptors can also activate JNK or NF-κB in some cells, although the precise mechanisms are less well defined.

DEATH RECEPTOR SIGNALING CAN ENGAGE THE MITOCHONDRIAL PATHWAY OF APOPTOSIS

In Chapter 5, we discussed how the BH3-only protein, BID, is a sensor for proteases, because cleavage of the protein can activate it to engage the mitochondrial pathway of apoptosis by activating BAX and BAK. When death receptors are ligated and activate caspase-8, this causes MOMP and engages the mitochondrial pathway of apoptosis. This is because caspase-8 is very efficient at cleaving BID and activating this BH3-only protein (Fig. 6.11).

In some cells, inhibition of MOMP by anti-apoptotic BCL-2 proteins has little or no effect on death receptor–mediated apoptosis. These have been called type I cells

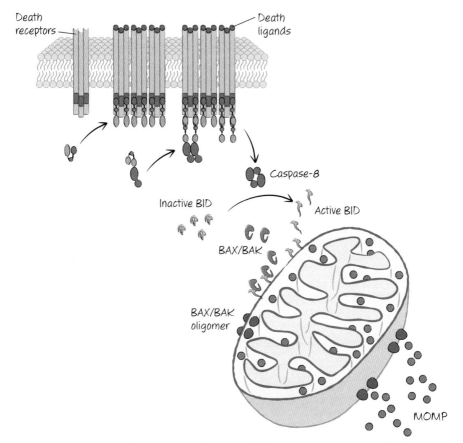

Figure 6.11. Death receptor signaling to MOMP via caspase-8 cleavage of BID.

in terms of their response to death ligands.[4] Lymphocytes and lymphoid tumors are examples of type I cells. In other cells, inhibition of MOMP effectively blocks apoptosis, and these are referred to as type II cells. Hepatocytes are an example of type II cells, as we will see.

The critical difference between type I and type II cells in the response to death receptor ligation may be the expression of XIAP. This protein blocks the activities of caspase-9 and the executioner caspase-3 and -7 (see Chapter 3) but does not inhibit caspase-8. When MOMP is triggered by cleaved BID, antagonists of XIAP, such as

---

[4] It is unfortunate (and confusing) that the forms of cell death are referred to as "type 1," "type II," and "type III," whereas the cells that respond differently to death receptor signaling are also referred to as "type 1" and "type 2" cells. In this chapter, any reference to "type 1 cells" or "type II cells" is to the latter.

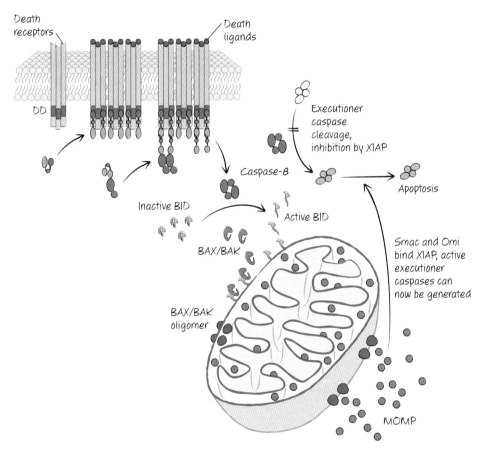

Figure 6.12. Death receptor–induced apoptosis in type II cells. BID-induced MOMP can also engage apoptosome formation and caspase activation, but this does not appear to be necessary for apoptosis by this mechanism.

Smac and Omi, are released from the mitochondrial intermembrane space and disrupt XIAP function, allowing apoptosis to proceed. This effect can be seen in cells lacking APAF1 (which therefore cannot activate caspases as a result of cytochrome c release), and this tells us that the function of MOMP to block XIAP can be critical for caspase-8 to activate caspase-3 and -7 and promote apoptosis by this pathway. This scheme is illustrated in Figure 6.12.

One example is the cells of the liver, the hepatocytes. Injection of mice with CD95-ligand or with antibodies that activate CD95 causes a rapidly lethal destruction of the liver. Expression of BCL-2 in hepatocytes prevents this lethal effect. Mice that lack BID are also resistant to these treatments. Therefore, these cells are type II cells that require engagement of the mitochondrial pathway following CD95 ligation to cause apoptosis. However, mice lacking both BID and XIAP are sensitive.

CD95 ligation on hepatocytes therefore engages caspase-8, but although caspase-8 can activate the executioner caspases, it seems that XIAP blocks them and preserves survival of both the cells and the animals. However, when caspase-8 cleaves and activates BID, the resulting MOMP releases proteins such as Smac and Omi that neutralize XIAP, and death ensues. At least in the case of hepatocytes, it is the expression and function of XIAP that make these type II cells; MOMP must be engaged to inhibit XIAP in order for apoptosis to proceed.

There may be another way in which some death receptors engage apoptosis. As mentioned above, signaling from the TNFR1 and TRAIL receptors can activate JNK. As we saw in Chapter 5, JNK can phosphorylate the BH3-only protein BIM, promoting its activity. Signaling from some death receptors might therefore trigger apoptosis in this way, and there is evidence to support this idea. If this is the case, we have to reconcile this with the finding that caspase-8 is required for apoptosis induced by death receptors. Unfortunately, at present, we know of no way for caspase-8 to participate in the activation of JNK or BIM.

DEATH RECEPTORS IN OTHER ANIMALS

Although there is a TNFR-family protein in *Drosophila*, its ligation does not appear to activate caspases or apoptosis in any situation. *Drosophila* also have a homolog of FADD (called dFADD) and a homolog of caspase-8 containing DEDs (Dredd). If these proteins are not involved in apoptosis, what do they do?

It may be that the role of this pathway in *Drosophila* (and presumably other insects) is in host defense. Dredd and dFADD participate in the production of antibacterial peptides in the fly. Intriguingly, this is because Dredd cleaves and thereby activates a protein related to NF-κB called Relish. This has led to the idea that caspase-8 in mammals may have a similar role in an NF-κB pathway, but currently this is controversial.

CHAPTER 7

Other Caspase Activation Platforms

OTHER WAYS TO APOPTOSIS

Although most apoptosis in mammals is by the mitochondrial pathway or death receptor ligation, caspase activation can occur by other mechanisms as well. Some of these lead to cell death and some have other functions.

Apart from the executioner caspases, caspases are activated by induced proximity. As we have seen, this allows the formation of dimers, which is generally caused by the binding of one or more adapter proteins to the prodomains of the caspase monomers. These adapter proteins help to define the caspase activation pathway involved. In this chapter, we discuss the activation and functions of two caspases that can participate in cell death but seem to have other functions as well: caspase-1 and -2. These caspases are not activated by the pathways that we have described in Chapters 4 and 6, but rather by other activation platforms.

CASPASE-1 IS ACTIVATED BY INFLAMMASOMES

As outlined in Chapter 3, caspase-1 is involved in the processing of interleukin-1β and interleukin-18, secretion of these and other proteins, and cell death in some settings. This form of cell death resembles apoptosis and is sometimes referred to as pyroptosis. The activation of caspase-1 occurs in complexes called *inflammasomes*, and unlike the other caspase activation platforms that we have discussed (the apoptosome and the DISC for caspase-9 and -8, respectively), inflammasomes can be composed of different adapter molecules. In most inflammasomes, however, a common feature is a molecule that binds to the CARD in the prodomain of a caspase-1 monomer (Fig. 7.1).

Caspase-1 is expressed in macrophages, dendritic cells, and some other cell types, and inflammasomes that activate this caspase assemble in these cells in response to a

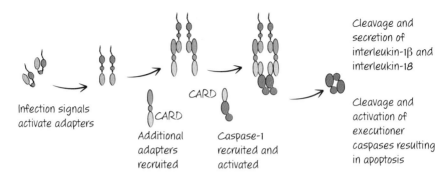

Figure 7.1. The basic inflammasome and activation of caspase-1.

wide range of signals, most of which are associated with infectious agents. These signals can be proteins, lipids, DNA, or RNA from pathogens, collectively referred to as pathogen-associated molecular patterns (PAMPs). Some inorganic materials, such as some crystals, also induce inflammasome formation, as do (probably) some materials released from necrotic cells. Another general term is sometimes used for these: damage-associated molecular patterns (DAMPs).[1] How PAMPs/DAMPs trigger inflammasome formation and how different inflammasomes are engaged are at the heart of a rapidly emerging area of inflammation research.

SEVERAL INFLAMMASOMES FOR CASPASE-1 ACTIVATION INVOLVE ASC

A small CARD-containing protein called ASC (apoptotic speck-forming CARD) can oligomerize and bind caspase-1 via CARD–CARD interaction to activate the protease by induced proximity. In addition to its CARD domain, ASC contains another death fold, a pyrin domain (PyD). ASC appears to oligomerize spontaneously to form large aggregates (appearing as specks in the cytoplasm), but this is inhibited by the levels of potassium ions normally found in cells. However, if potassium channels in the plasma membrane allow a sufficient efflux of potassium, this can allow ASC to oligomerize and caspase-1 is activated. Any prointerleukins present can be processed and secreted. If potassium levels fall severely, the extent of caspase-1 activation can be sufficient to cleave and activate executioner caspases and kill the cell (Fig. 7.2).

One such potassium channel is $P2X_7$. This is activated by extracellular ATP, which might act as a damage signal that cells (pathogens or host cells) are lysing in the vicinity.

---

[1] DAMPs are sometimes called danger-associated molecular patterns, based on the danger hypothesis. Because "danger" is rather tautological, defined as anything that elicits a response, we prefer the term "damage," but this has its own problems.

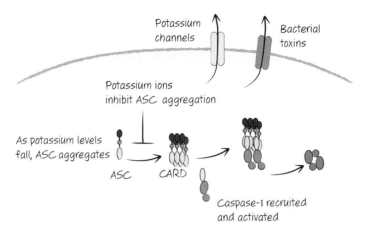

Figure 7.2. Potassium efflux can promote caspase-1 activation via ASC.

Often, however, bacteria cause potassium efflux directly by virtue of pore-forming toxins that they produce.

TLRs INDUCE OTHER INFLAMMASOME COMPONENTS AND CASPASE-1 SUBSTRATES

A set of receptors related to the *Drosophila* Toll protein are involved in recognizing many PAMPs. These Toll-like receptors (TLRs) recognize components of bacterial cell walls, bacterial RNA, and fungal components. When they interact with PAMPs, TLRs activate NF-κB, which in turn induces the interleukin targets of caspase-1 as well as additional inflammasome components that can bind to and oligomerize ASC to facilitate caspase-1 activation. In most cases, however, engaging a TLR is not sufficient to trigger the inflammasome unless potassium levels in the cell decrease. We will see, however, that there are other ways to induce the activation of caspase-1. In Figure 7.3, some of the TLRs in human cells are illustrated, with some of the PAMPs they recognize.

NLRs PARTICIPATE IN INFLAMMASOMES

In addition to the TLRs, a large set of intracellular molecules that appear to recognize PAMPs and DAMPs are the Nod-like receptors (NLRs, named for the first one found, Nod-1). The NLRs are found throughout the animal kingdom and even in plants, where they are involved in host defense against infection. Collectively, TLRs and NLRs are often referred to as pattern-recognition receptors (PRRs).

Several NLRs contain death folds, including CARDs and PyDs, and some of these have been shown to participate in the creation of inflammasomes. The most important

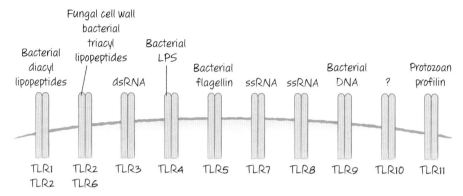

Figure 7.3. Some TLRs and their PAMPs. Some TLRs (e.g., TLR9) are not found on the cell surface, despite their placement in the figure.

of these known are NLRP3 (also called NALP3 and cryopyrin) and IPAF. All of the NLRs (including those in plants) also have a NACHT domain (see Chapter 4) and a long tail. In many NLRs, this tail is a leucine-rich region (LRR) believed to be specific for a type of PAMP or DAMP that appears in the cell. Some of the mammalian NLRs and what they appear to recognize are listed in Figure 7.4.

Whether NLRs actually respond directly to a DAMP or PAMP has not been formally proven. It remains possible (and in some cases likely) that other molecules participate in this recognition event.

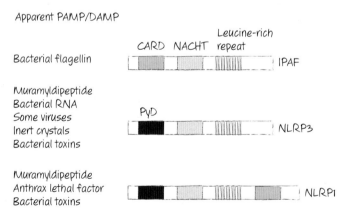

Figure 7.4. Some NLRs that function in inflammasomes. PAMPs and DAMPs that induce them are listed. Bacterial toxins may function to alter potassium levels, thereby indirectly facilitating inflammasome formation.

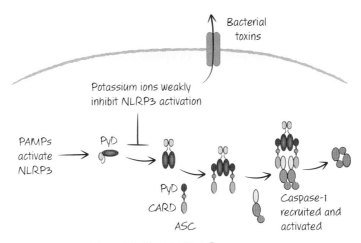

Figure 7.5. The NLRP3 inflammasome.

THE NLRP3 INFLAMMASOME

NLRP3 contains a PyD that binds to the PyD of ASC. In addition, NLRP3 has a NACHT domain and an LRR. NLRP3 and ASC are required for the formation of inflammasomes in response to a number of PAMPs, including those from *Staphylococcus aureus*, *Lysteria monocytogenes*, and bacterial RNAs. This does not depend on TLR signaling. However, TLR signals induce the expression of NLRP3 to increase sensitivity. The NLRP3-ASC inflammasome is also sensitive to potassium levels (but less so than ASC alone), and efflux of this ion potentiates NLRP3 effects (Fig. 7.5).

Some noninfectious materials also act to engage the NLRP3 inflammasome to activate caspase-1. These include asbestos, crystals of uric acid (the cause of gout), and crystals of calcium pyrophosphate dihydrate (CPPD, the cause of pseudogout) (Fig. 7.6).

THE IPAF INFLAMMASOME

Another important NLR for inflammasome formation is IPAF (also called CARD12 and NLRC4), which is required for the response to bacterial flagellin from *Salmonella typhimurium* and *Legionella pneumophila* and to *Shigella flexieri* infection. In the case of bacterial flagellin, the response requires the presence of the protein in the cytoplasm of the infected cell, which occurs as a consequence of its secretion by the bacterium.

IPAF has a CARD. Following activation, this interacts with ASC through a CARD–CARD interaction that does not preclude the CARD–CARD interaction between ASC and caspase-1. Alternatively, it is possible that the CARD of IPAF binds directly to the CARD of caspase-1 (Fig. 7.7). Therefore, if IPAF dimerizes, it can bring together and activate two caspase-1 monomers, through direct or indirect interactions.

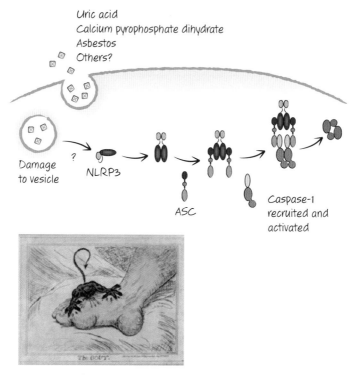

Figure 7.6. Inert crystals induce the NLRP3 inflammasome in gout and other diseases.

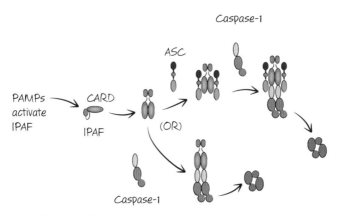

Figure 7.7. Two possible IPAF inflammasomes. IPAF may or may not require ASC to activate caspase-1, and the possible interactions are shown. They may both be correct under different circumstances. The role of NAIP5 is not shown.

Another NLR, called NAIP5, appears to be important in the IPAF response to *Legionella*. NAIP5 does not have a death fold. Instead, it has a region with three BIR domains. It therefore resembles an IAP protein (see Chapter 3) but without caspase-inhibitory activity (as far as we know). How NAIP5 cooperates with IPAF to activate caspase-1 is not clear.

OTHER INFLAMMASOMES

The lethal toxin of anthrax activates caspase-1 via another inflammasome, which involves the NLR NLRP1b (also called NALP1b). NLRP1b is different in humans and rodents, and the way in which it activates caspase-1 is somewhat controversial. It has a CARD, and although some studies indicate that the activation of caspase-1 by NLRP1 requires ASC, others show that this protein can interact directly with caspase-1 via a CARD–CARD interaction. These two possibilities are shown in Figure 7.8.

Another inflammasome is composed of yet another NLR, NLRP2 (also called NALP2), and ASC. It is not known what NLRP2 recognizes to trigger caspase-1 activation.

There is a type of inflammasome that does not involve an NLR. Double-stranded DNA in the cytosol binds to a protein called AIM2. AIM2 has a PyD that binds to the PyD of ASC. Therefore, when DNA appears in the cytosol, clusters of AIM2 form and, in turn, activate caspase-1 (Fig. 7.9).

AIM2 might be a sensor for infection by DNA viruses, although it also binds to any other double-stranded DNA, including that from the host (provided it appears in the cytosol). For this reason, the AIM2 inflammasome is thought to have a role in the autoimmune response to double-stranded DNA in systemic lupus erythematosis and related diseases.

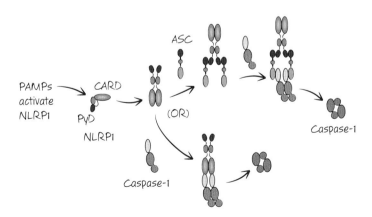

Figure 7.8. Two models of the NLRP1 inflammasome. Both may be correct in different settings.

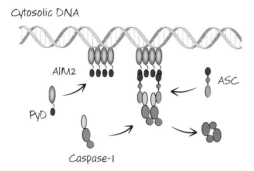

Figure 7.9. The AIM2 inflammasome.

CELL DEATH BY CASPASE-1

As we have discussed, the activation of caspase-1 can lead to apoptotic cell death, but it is not clear that this is always the case (unfortunately, it is also not clear that it is *not* always the case). Presumably, if caspase-1 is activated at low levels, this allows it to effect interleukin processing and secretion without causing cell death, but this has not been proven. If this is the case, however, we can speculate that the levels of potassium in the cell influence this decision to die or not. In addition, there are a number of proteins that affect caspase-1 activation. For example, caspase-12 (see Chapter 3) appears to competitively inhibit caspase-1, and there are other molecules with similar modulating activity.

Caspase-1 causes cell death by cleaving and thereby activating executioner caspases, caspase-7 being more important than caspase-3 in this case. It can also activate BID and thereby engage the mitochondrial pathway of apoptosis (see Chapter 5). Bacteria that infect cells often induce cell death via activation of caspase-1, and in several cases the inflammasome responsible for this effect has been identified (as indicated above). For example, *Shigella flexieri* kills macrophages in a manner that depends on IPAF, ASC, and caspase-1.

NLR ACTIVATION AND THE RETURN OF THE *JUST SO STORY*

The similarity between APAF1 and the NLRs hopefully has not escaped the reader, and indeed we can include it among the NLRs on the basis of sequence similarity (Fig. 7.10).

This may mean that we can draw an analogy with APAF1 to explain how NLRs are activated by a ligand. If the LRR regions act like the WD region of APAF1, binding to a ligand may cause conformational changes that expose the nucleotide-binding site in the NACHT domain. Binding of a nucleotide would then, as in APAF1, induce further conformational changes to expose an oligomerization domain and the protein–protein interaction site (e.g., CARD or PyD) to interact with ASC. At present, however, this is all speculation.

OTHER CASPASE ACTIVATION PLATFORMS   107

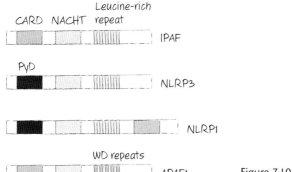

Figure 7.10. APAF1 is in the NLR family.

The relationship between APAF1 and the NLRs is tantalizing and brings us back to our *Just So Story* about the evolution of apoptosis. It is easy to imagine that the ancient cell infected by the "mitochondrion to be" had a defense system in place to recognize such infections, using some type of NLR. Perhaps this NLR used bacterial cytochrome c as a PAMP, because this protein did not exist in the cell at that time. The recognition of the PAMP could have engaged a cell death mechanism to prevent the spread of the parasite to other cells. In this fantasy, the NLR eventually became APAF1, and the cell death mechanism became the mitochondrial pathway of apoptosis (Fig. 7.11).

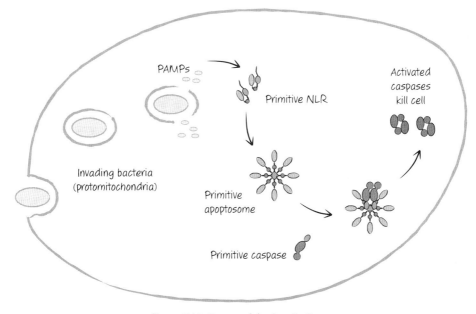

Figure 7.11. Return of the *Just So Story*.

THE ACTIVATION OF CASPASE-2

Caspase-2 is a bit of a puzzle. It is the most highly conserved of the caspases throughout the animals and has been implicated in cell death caused by heat shock, cytoskeletal disruption, metabolic perturbation, and perhaps DNA damage, among other situations. However, in none of these cases has caspase-2 been unambiguously shown to be *required* for cell death.

Caspase-2 has a long prodomain containing a CARD and is activated by induced proximity (see Chapter 3). Unlike the other caspases, however, it is a rather poor activator of executioner caspases. Instead, it may require BID to cause cell death by engaging the mitochondrial pathway of apoptosis.

These considerations raise an intriguing question: Is the primary function of caspase-2 to cause apoptosis? Caspase-2 may have another function in cells that has not been identified. But it does seem to be important. Mice lacking caspase-2 are developmentally normal but age prematurely and are more prone to cancer in at least one model system.

The activation of caspase-2 involves an adapter protein called RAIDD, which binds to caspase-2 by a CARD–CARD interaction. RAIDD does not appear to oligomerize itself, but instead may require additional molecules. One such molecule is PIDD. PIDD has a DD, and it binds to RAIDD via a DD–DD interaction. This is illustrated in Figure 7.12.

The caspase-2 activation platform is called a PIDDosome. The PIDDosome has been resolved at a structural level, which shows it to have multiple PIDD and RAIDD subunits (Fig. 7.13). This is similar to the structure of the CD95-DD–FADD-DD structure involved in the activation of caspase-8 (see Chapter 6).

PIDD is an interesting protein that undergoes extensive processing. It has a larger form that does not seem to be involved in caspase-2 activation, but instead activates the transcription factor NF-κB. However, this form of PIDD processes itself to a smaller form that can engage RAIDD. This processing is by an intein mechanism, in which sequences in the protein interact in a manner very similar to the function of proteases.

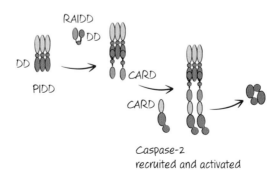

Figure 7.12. The PIDDosome activates caspase-2.

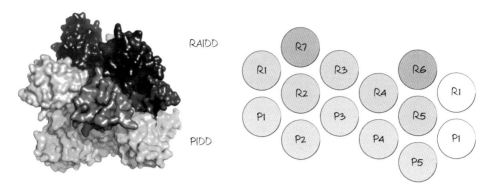

Figure 7.13. PIDDosome structure. The interactions of PIDD (P) and RAIDD (R) DDs.

This results in autocleavage of the protein to its smaller form. How the relative levels of the different species of PIDD are controlled is not clear.

This dual activity of PIDD is reminiscent of the ability of some of the death receptors to activate caspase-8 (and apoptosis) as well as NF-κB. It is not known if NF-κB, induced by PIDD, can induce the expression of genes that influence the function of caspase-2, but the idea is intriguing.

repair enzyme poly-ADP-ribose polymerase (PARP) engages processes that consume a large amount of NADH, which is produced by metabolic pathways. This enzyme can therefore promote necrotic cell death in some circumstances by using up NADH, especially if nutrients are limited.

DNA damage can also trigger apoptosis through signaling mechanisms, including the action of p53 (see Chapter 11). However, if DNA damage is very extensive, this can result in a form of necrosis[2] that can be blocked by pharmacological inhibitors of PARP or additional sources of NADH. For example, the addition of nicotinamide can reduce necrosis induced by high-dose γ radiation.[3]

Reactive oxygen species can damage DNA, and high levels of these can cause cell death that depends on PARP. Mice lacking PARP develop normally but are resistant to neuronal damage induced by hydrogen peroxide. We will see below other ways in which reactive oxygen species that lead to necrosis can be generated.

EXCITOTOXICITY

Neurons can undergo active necrosis in response to high levels of glutamate, a neurotransmitter in the brain. This can occur as a consequence of ischemic injury (i.e., stroke) in the brain (discussed in detail below). Glutamate-induced neuronal necrosis is often referred to as *excitotoxicity* or excitotoxic death.

When neurons are exposed to high levels of glutamate, this causes an influx of calcium ions into the cells. This, in turn, activates a complex enzyme, NADPH oxidase, that produces reactive oxygen species. Originally, NADPH oxidase was found in neutrophils, which use it to destroy bacteria and other pathogens. But we now realize that many cells in the body express this or a related form of this enzyme, which is more simply referred to as Nox.

Nox produces reactive oxygen species by moving an electron from NADPH to oxygen to make superoxide. But, unlike the electron transport chain of mitochondria, this does not generate energy but instead produces protons. These must be transported out of the cell, which generates a high charge difference across the plasma membrane. This charge is seen as excitation in glutamate-exposed neurons. As a result of the superoxide, PARP is activated, and NADH is depleted (Fig. 8.3). The loss of energy in the cells promotes their death by necrosis. Inhibitors of Nox prevent both PARP activation and death by excitotoxicity.

---

[2] It may be confusing to suggest that apoptosis is superseded by necrosis under some circumstances, and in real life, the effects can be mixed. However, as we noted, the mechanisms are the key, and blocking one mechanism (e.g., apoptosis) will not block the other (e.g., necrosis).

[3] Nicotinamide has multiple roles in cells, and it may be simplistic to conclude that its effect on cell death is only as a precursor for NADH.

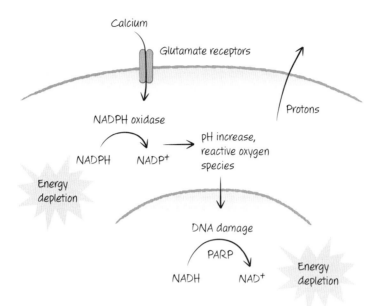

Figure 8.3. One view of excitotoxicity. Calcium influx triggers NADPH oxidase, which produces reactive oxygen species (damaging DNA and inducing PARP) and hydrogen ions (creating plasma membrane charge). Necrosis occurs as a consequence of energy depletion.

ISCHEMIA/REPERFUSION INJURY

Ischemic injury occurs when the blood supply to a tissue, such as the heart or brain (or other tissues), is disrupted. Not surprisingly, deprivation of oxygen and nutrients can cause necrosis. However, when the blood supply is restored (reperfusion) this often causes a wave of cell death that can have catastrophic consequences. In the heart, for example, the extent of cell death is directly linked to remodeling events that over time cause heart enlargement, accompanied by thinning of the muscle walls and ultimately heart failure. This is why heart attacks so often predispose individuals to subsequent heart attacks. Controlling the extent of damage from ischemia/reperfusion injury is a major therapeutic goal.

Why does ischemia/reperfusion cause so much damage? Part of the answer relates to changes in potassium levels in the cells. Following reperfusion, potassium channels open in the plasma membrane, causing a drop in intracellular potassium that triggers a range of changes, including a rise in intracellular calcium. As we have seen, this can activate Nox. In addition, the increase in calcium can also activate the protease calpain, which at high levels can cause necrosis (Fig. 8.4). Animals that are treated with inhibitors of calpain, or lack of this enzyme, show reduced damage because of ischemia/reperfusion injury.

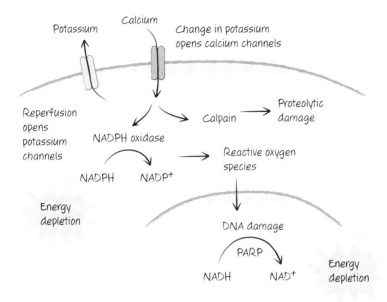

Figure 8.4. One view of ischemia/reperfusion injury. Reperfusion induces an opening in potassium channels, and in turn, the change in potassium opens calcium channels. The influx of calcium triggers NADPH oxidase, and its consequences, and activates calpain, a protease. Death may be a consequence of energy depletion, calpain action, or both.

MITOCHONDRIA IN ISCHEMIA/REPERFUSION INJURY

Mitochondria are important in apoptosis (see Chapters 4 and 5), but they can also have an important role in necrosis. As we discussed in Chapter 4, mitochondria can respond to high levels of calcium, reactive oxygen, and other signals by opening a channel in the inner membrane, the permeability transition pore (PTP). This causes the mitochondrial matrix of the organelle to swell, eventually rupturing its membranes—the so-called mitochondrial permeability transition (MPT). One protein that we saw to be clearly involved is the peptidylproline isomerase cyclophilin D. Mice lacking cyclophilin D are developmentally normal, but mitochondria from these mice show defective PTPs.[4] These mice are strikingly resistant to ischemia/reperfusion injury of the heart and brain (Fig. 8.5). The simplest explanation for this is that the MPT has a role in the damage. MPT can contribute to necrosis by both destroying mitochondrial energy generation and ablating the ability of mitochondria to scavenge reactive oxygen.

Cyclosporin A, a drug that inhibits cyclophilins including cyclophilin D, limits ischemia/reperfusion injury in rodents and can reduce heart damage in humans. It is

---

[4] Remember that cells from these animals display normal apoptosis, and we therefore suspect that this is not an important mechanism for MOMP in apoptosis (see Chapter 4).

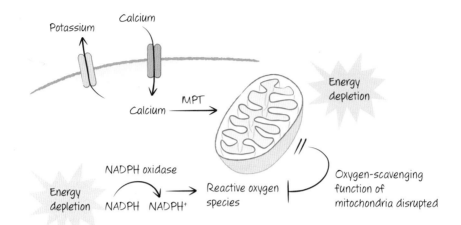

Figure 8.5. MPT in ischemia/reperfusion injury. Elevated calcium induces the mitochondrial permeability transition (MPT), which disrupts energy generation and curtails reactive oxygen scavenging. Mice lacking the MPT are resistant to ischemia/reperfusion injury.

tempting to think that this is because the drug blocks the MPT. However, it might not be so simple. Cyclosporin A undergoes a conformational change when it binds to cyclophilins, and the complex of the drug and cyclophilins in the cytoplasm is a potent inhibitor of the enzyme calcineurin, and this effect may be more important.[5]

The damage in ischemia/reperfusion injury is not limited to necrosis; the area around the initial damage undergoes extensive apoptosis. This might be because of signals such as reactive oxygen species released from the necrotic cells or the effects of inflammatory cells that are activated by the necrosis (discussed in more detail in Chapter 10). Inflammatory cells also produce ligands for death receptors (see Chapter 6). Mice lacking one of the death receptors, CD95, show reduced injury following ischemia/reperfusion. As we will see, however, signaling from death receptors can also induce necrosis.

DEATH RECEPTORS CAN CAUSE NECROSIS

As we saw in Chapter 6, death receptors of the TNFR family can trigger an apoptotic pathway. However, in some cases, they can also induce necrosis, although this form of cell death has some features of type II (autophagic) cell death (see below). For this reason, the necrotic form of cell death induced by death receptors is sometimes referred to as programmed necrosis or necroptosis.

---

[5] Calcineurin is important for the activation of a set of transcription factors called NFAT (it is the inhibition of NFAT activation that accounts for the ability of cyclosporin A to block T lymphocyte activation and tissue rejection, for which it is used). NFAT also has important roles in blood vessel development and other functions, and at this point, we simply do not know how much these contribute to the beneficial effects of the drug in ischemia/reperfusion injury.

When TNF engages its death receptor (TNFR1), this activates a TRADD-TRAF-RIPK1 complex that dissociates from the receptor and recruits the adapter FADD and caspase-8 (see Chapter 6), and apoptosis proceeds. However, in many cells, if caspase-8 is inhibited, death nevertheless ensues. This death is necrotic and requires RIPK1. Unlike the activation of NF-κB by this receptor (see Chapter 6), however, this programmed necrosis requires RIPK1 kinase activity.

Inhibitors called *necrostatins* block the kinase activity of RIPK1 and prevent necrosis induced by TNF. They can also reduce cell death caused by ischemia/reperfusion injury and pathological cell death in other settings, although in these cases whether this is caused by TNF, another death ligand, or other signals that engage RIPK1 is not known.

RIPK1 appears to mediate TNF-induced necrosis in two ways (that may be interrelated). First, it can recruit and activate an NADPH oxidase related to Nox[6] that generates superoxide. Inhibitors of Nox reduce necrosis induced by TNF. Second, RIPK1 also recruits and activates another kinase, RIPK3, that is also required for TNF-induced necrosis. Cells lacking RIPK1 or RIPK3 do not die by programmed necrosis, and if caspase-8 is inhibited, the cells survive TNF treatment.

RIPK3 activates a number of enzymes, including several that are involved in metabolism. These enzymes increase the use of glucose and glutamine in the cell, and their activity appears to be important for the ensuing necrosis. How these contribute to cell death is far from clear. What we know of the mechanism of TNF-induced necrosis is illustrated in Figure 8.6.

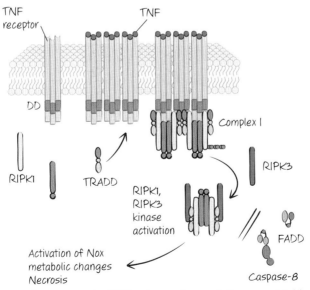

Figure 8.6. TNFR signaling engages the RIPK pathway of necrosis if caspase-8 is blocked.

---

[6] This is Nox1, a relative of the Nox that we have been discussing.

This is a very active area of research, and by the time you read this, we may know more about this intriguing pathway. Besides TNFR1, other death receptors appear to be able to engage RIPK1-mediated necrosis, and there are probably other ways in which cells can activate RIPK1 kinase.

Why do cells have this backup mechanism for cell death? One likely answer relates to viruses. As we mentioned, viruses deploy strategies to inhibit caspase-8 as a way to block apoptosis of host cells that is induced by cells of the immune system. When this occurs, the production of reactive oxygen species and the destruction of the cell by necrosis is a good alternative to combat the infection.

CASPASE-8 AND FLIP INHIBIT PROGRAMMED NECROSIS

Inhibition of caspase-8 allows TNF to induce necrosis rather than apoptosis, and therefore caspase-8 would appear to inhibit this pathway. Indeed, caspase-8 cleaves RIPK1 and RIPK3, preventing their activity, and this may be how caspase-8 blocks necrosis.

But if activation of caspase-8 commits the cell to die by apoptosis anyway, can inhibition of RIP-dependent necrosis preserve the cell? When TNF binds to TNFR1, it induces the activation of NF-κB, and this transcription factor induces the expression of FLIP, which, as we have seen, is a protein that is related to the caspases but lacks a catalytic site. Like caspase-8, FLIP can be recruited to the DISC in the death receptor pathway. At the DISC, FLIP forms a complex with caspase-8 that is proteolytically active but unstable because caspase-8 is not cleaved; consequently, FLIP prevents apoptosis. However, this complex may also prevent programmed necrosis. One possibility is that the caspase-8–FLIP dimer might cleave RIPK1 in the complex and thereby ensure that the cell does not undergo apoptosis nor necrosis and therefore survives (Fig. 8.7).

Mice lacking caspase-8 or FLIP die during development because of extensive cell death in the endothelium. It is very tempting to imagine that this is because the caspase-8–FLIP dimer is required to protect the cells from death triggered by RIPK1.

AUTOPHAGIC (TYPE II) CELL DEATH

Autophagic cell death is characterized by the presence of large vacuoles in the cytoplasm, as well as molecular markers of autophagy (see below). In general, the characteristic features of apoptosis, including plasma membrane blebbing, nuclear condensation, and chromatin fragmentation, are not seen, and caspases do not have a role.

Autophagic cell death may be a victim of its name; for the most part, neither autophagy (primarily a survival mechanism) nor molecular components of the autophagy pathway are responsible for autophagic cell death. There are exceptions and some tantalizing observations, as we will see. But most instances of autophagic cell death appear to be accompanied by autophagy, rather than being caused by it. In fact, autophagy appears to antagonize cell death, and inhibition of autophagy can dramati-

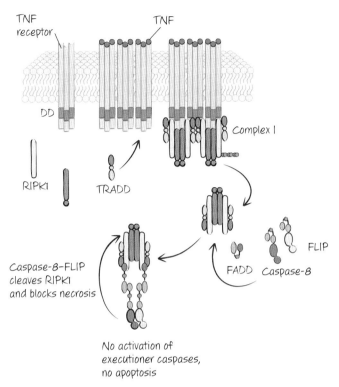

Figure 8.7. Caspase-8–FLIP prevents death. Shown is one way that caspase-8–FLIP heterodimers may block both apoptosis and necrosis. The cleavage of RIPK1 by caspase-9–FLIP is speculation.

cally increase it. This is a controversial area and there are sure to be important revelations in the coming years.

AUTOPHAGY AS A SURVIVAL MECHANISM

In general, autophagy is considered a survival mechanism for cells. Formally, there are three types, but here we are only concerned with macroautophagy.[7] This is usually referred to simply as autophagy and we use this nomenclature here.

Autophagy is found throughout the eukaryotes and there is remarkable conservation of the components of the central pathway. It provides energy when external nutrients are depleted or the pathways for taking them up are disrupted, removes excess or damaged organelles and other cellular components, and is involved in the degradation of long-lived proteins and protein aggregates in cells. Autophagy also functions in cellular defense to isolate and destroy invading organisms.

---

[7] The other two types are called microautophagy and chaperon-mediated autophagy.

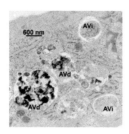

Figure 8.8. Autophagosomes. Shown are early autophagosomes (AVi) and autophagolysosomes (AVd).

Because autophagy is an important survival mechanism, it is useful to go into some detail about how it works. This will also help us to understand how it might be engaged to kill cells as well.

THE AUTOPHAGY PATHWAY

Autophagy involves the generation of membrane vesicles in the cell, called autophagosomes, that enclose cytoplasm, organelles, protein aggregates, or invading organisms and carry them to lysosomes, with which the autophagosomes fuse. The degradative enzymes in the lysosomes break down the contents that are then reused as sources of energy and raw materials by the cell (Fig. 8.8).

Autophagy in mammals occurs by the hierarchical action of four complexes: ATG13, Beclin-1–phosphoinositide-3-kinase (PI3K), ATG5–ATG12, and LC3. Each complex recruits the next,[8] ultimately building a double membrane structure called a phagophore. The phagophore closes to engulf cytoplasm, the complexes dissociate, and the result is an autophagosome (Fig. 8.9).

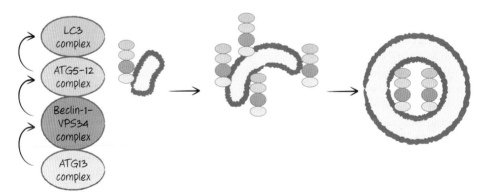

Figure 8.9. Simplified, hierarchical autophagy pathway.

---

[8] This is only a model; for example, it is not certain that ATG13 recruits the Beclin-1–PI3K complex. However, the model is useful for understanding how autophagy may, in principle, work, and there is evidence to support it.

MAMMALIAN AUTOPHAGY IN DETAIL

Autophagy in mammals is initiated by the ATG13 complex.[9] This contains a kinase, ULK1, ATG13, and a protein called FIP200. When autophagy is activated by starvation or cellular damage, ATG13 dissociates and recruits the next complex in the pathway. This complex contains Beclin-1 (see Chapter 5) and a class III PI3K called VPS34, together with other components. This generates the lipid, phosphatidylinositol 3-phosphate (PI3P), that recruits the next components of the autophagy pathway by a mechanism that is obscure in mammals.[10]

ATG7, an enzyme similar to a ubiquitin E1-ligase, is now brought in, binding a small protein called ATG12 and passing it to an E2-like enzyme, ATG10, which then places it on another protein, ATG5. The ATG5–ATG12 complex recruits another protein, ATG16, and this forms the platform for the next step in the process (Fig. 8.10).

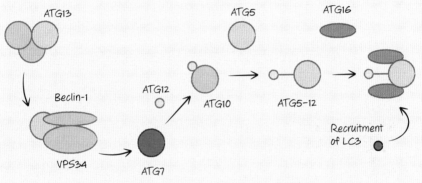

Figure 8.10. The initial steps in autophagy.

The ATG5–ATG12–ATG16 complex recruits another protein, LC3, and ATG7 again acts as a ligase, passing it to ATG3, which then places LC3 onto a lipid, phosphatidylethanolamine (PE).[11] This LC3–PE complex is the building block of the phagophore, which begins to grow (Fig. 8.11). When it is complete, the other proteins uncoat, and it is now an autophagosome, ready to fuse with a lysosome.

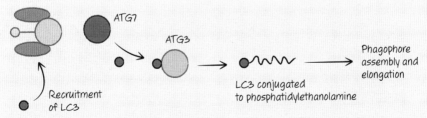

Figure 8.11. LC3 and the phagophore.

---

[9] Also called the ULK1 complex, because ULK1 is the catalytic subunit.

[10] In yeast, a protein called ATG18 binds to PI3P, and this acts to recruit the other players. There is a mammalian counterpart called WIPI-1a that might have this role in mammalian autophagy, but this is not yet clear.

[11] Another view holds that the ATG5–12 pathway and the LC3–PE pathways are parallel, both contributing to the construction of the phagophore without the hierarchical organization proposed here.

ENGAGING THE AUTOPHAGY PATHWAY

Autophagy is inhibited in several ways, and how this inhibition is itself regulated provides some clues as to how autophagy is engaged in the cell. Although we consider two mechanisms for induction of autophagy, it may be worth noting that there are almost certainly other ways in which this pathway can be triggered.

One key regulator of autophagy is also a key regulator of cell growth, metabolism, and protein synthesis, a kinase called mTOR. It is found in two different protein complexes, TORC1 and TORC2, but it appears to be TORC1 that controls autophagy. When nutrients are plentiful, mTOR is active; cells grow and autophagy is inhibited. However, when energy levels are low, AMP accumulates, and this activates a kinase, AMPK, that inhibits mTOR, and autophagy is engaged. Similarly, growth factor receptor signaling activates AKT (see Chapter 5), and this inhibits a kinase, GSK3, that is an inhibitor of mTOR. When AKT is inactive, mTOR is inhibited, and autophagy proceeds (Fig. 8.12). The drug rapamycin is a potent inhibitor of mTOR (in fact, TOR means "target of rapamycin") and the addition of this drug to cells induces autophagy.

Another way in which autophagy is regulated involves BCL-2 (and possibly other anti-apoptotic BCL-2 proteins), which can bind and sequester Beclin-1 using the same BH groove used to sequester BH3-only proteins in apoptosis. Indeed, Beclin-1 has a BH3 domain, although this binds weakly to BCL-2 compared with others that we have discussed. It seems that if BH3-only proteins are expressed, they can neutralize the interaction between BCL-2 and Beclin-1, and if MOMP does not occur (i.e., BAX and BAK are not activated), autophagy can result.[12]

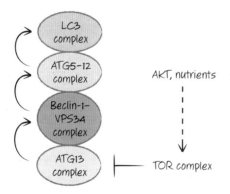

Figure 8.12. TOR and autophagy.

---

[12] In Chapter 5, we outlined one case in which this effect may be important. A BH3-only protein called NIX, which is not a potent potentiator of apoptosis, is important for the removal of mitochondria during the development of mammalian red blood cells. It is possible (although not formally proven) that NIX does this by disrupting the interaction between Beclin-1 and BCL-2. Another BH3-only protein, BNIP3 (which is also a poor inducer of apoptosis), is related to NIX, and this protein may also be involved in the autophagic removal of mitochondria in some settings.

NONAPOPTOTIC CELL DEATH PATHWAYS

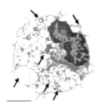

Figure 8.13. An autophagic survivor. (Arrows) Autophagosomes.

Autophagy can also be caused by DNA damage. This induces UVRAG, a protein that promotes the activity of VPS34 in the Beclin–PI3K complex (see above). Autophagy in this setting may provide additional energy needed for DNA repair.

In each of these scenarios, autophagy protects cells. If it is blocked, apoptosis ensues. However, when apoptosis is blocked and autophagy is active—for example, by removal of BAX and BAK—autophagy will keep the cells alive for extended periods, wearing them down to mere skeletons of their former selves (Fig. 8.13). If autophagy is also blocked, the cells die by necrosis.

WHAT IS AUTOPHAGIC CELL DEATH?

The above considerations help to clarify why a stressed cell that may undergo apoptosis engages the autophagic pathway as a result of the stress. Therefore, autophagy and apoptosis can occur in the same cell, but in such cases this is apoptosis and not autophagic cell death. However, cells often die without displaying features of apoptosis, instead displaying features of autophagy. In such cases, this is autophagic cell death (type II cell death), and it is seen in the response of tumor cells to several therapeutic agents. Usually, this is cell death that is associated with but not caused by autophagy, and if autophagy is blocked, the cells die by necrosis. Most autophagic cell death has this feature.[13]

WHEN AUTOPHAGY KILLS

Nevertheless, in some cases, it appears that autophagy can indeed promote (type II) cell death. The best-studied case is in *Drosophila* metamorphosis (discussed in Chapter 11). Larvae have very large salivary glands that die during metamorphosis. Some of this death is by apoptosis, but most of it has the form of type II cell death. If autophagy genes are defective in the developing fly, much of the cell death does not occur.

In mammals, there are several examples wherein type II cell death appears to be promoted by autophagy.[14] One way that this can occur is if the autophagic process is

---

[13] This may change as more systems are explored in vivo, and new situations in which autophagy actually kills cells are identified.

[14] Often, this has been shown by knocking down levels of expression of key proteins with siRNA. This is not without problems, because siRNA can have off-target effects, including interferon responses and other unexpected consequences that could affect cell death. Here, though, we will assume that there is something to the idea that autophagy can promote cell death in mammals.

activated but blocked before the contents of the autophagosome are degraded in lysosomes. As a result, the cytoplasm is disrupted, but the cell does not gain nutrients from the process, leading to death.[15] Alternatively, it has been proposed that another way in which autophagy promotes cell death is by the removal of catalase, a long-lived protein that neutralizes some reactive oxygen species. We can also imagine that, if autophagy removes other key survival proteins as well, cell death may result.

There is another interesting possibility, albeit untested. As we have discussed, autophagy involves fusion of the autophagosome with lysosomes in the cell. It is conceivable that if the process accelerates, so that incompletely formed autophagosomes engage lysosomes, the contents of the latter may be released into the cytosol. This would result in cell death as the destructive contents of the lysosome set to work. As mentioned above, large vacuoles are often associated with autosomal cell death, and it is possible that they are produced by such an interaction. Support for this notion comes from studies on a protein called DRAM. DRAM is associated with lysosomal membranes and is induced in some pathways leading to cell death. In some cases, inhibition of the expression of DRAM can protect cells from what appears to be type II cell death.

Perhaps the most interesting possibility relates to programmed necrosis. As we discussed, cells lacking caspase-8 can die following exposure to TNF in a manner that depends on RIPK1 kinase activity. In some studies, inhibition of autophagy prevents this cell death. Similarly, T lymphocytes that lack caspase-8 or FADD fail to proliferate and die following activation, and this cell death is blocked (and proliferation restored) by inhibition of RIPK1 kinase. Intriguingly, there is also evidence that disruption of autophagy can also rescue these T cells. It appears that the ATG5–ATG12 complex in the autophagic pathway recruits the adapter, FADD. FADD, caspase-8, and FLIP, associated with ATG5–ATG12, might cleave and inhibit RIPK1 kinase to prevent cell death when autophagy is engaged. If caspase-8 is not active, RIPK1-mediated necrosis ensues. This intriguing scenario links at least some forms of autophagic cell death to programmed necrosis (Fig. 8.14).

Normally during autophagy, ATG5–12 dissociates from the membrane when the autophagosome forms. If the model shown in Figure 8.14 is correct, it may be that a determining factor in whether autophagy contributes to cell death is the continued presence of ATG5–12 on the autophagosome membrane. This is only speculation, however.

MITOTIC CATASTROPHE

During mitosis, the nuclear membrane dissolves and chromosomes are segregated to the poles of the dividing cell. If this process is disrupted, cells die, and this is often

---

[15] This can be done pharmacologically, by addition of an autophagy inducer such as rapamycin and a lysosome inhibitor such as chloroquin. This strategy is being explored as a therapeutic regimen for cancer. Some viruses also inhibit lysosomes, and the metabolic stress caused by infection may result in defective autophagy that kills the cell in this manner.

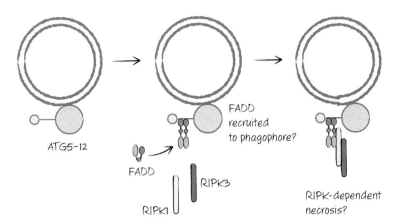

Figure 8.14. Hypothetical connection between autophagy and necrosis. Abundant ATG5–12 on the phagophore can bind FADD. This may recruit RIPK1 and RIPK3 to signal for necrosis.

called mitotic catastrophe. This can occur if the DNA is extensively damaged, the cell cycle machinery is stalled, or microtubules are dysfunctional.

Often, mitotic catastrophe results in apoptosis. How the mitotic machinery is linked to control of apoptosis is not clear, but one possible mechanism is intriguing. CDK1, the kinase that orchestrates mitosis, phosphorylates caspase-2 (see Chapter 7), preventing its activation. When mitosis is not completed on schedule, CDK1 activity declines and caspase-2 becomes active. Cells might therefore use caspase-2 to activate apoptosis when mitosis fails. There are likely other mechanisms as well.

If cells are faced with defective mitosis and apoptosis is blocked, cells die nevertheless, showing features of autophagic cell death or necrosis. An example of this is seen in the intestines of mice lacking BAX and BAK following irradiation (Fig. 8.15). How this cell death occurs remains obscure.

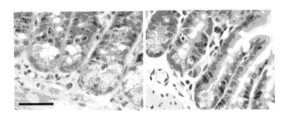

Figure 8.15. Irradiation causes nonapoptotic cell death in intestines lacking BAX and BAK. Mice with (*left*) or without (*right*) intestinal BAX and BAK were irradiated. Although apoptosis (asterisks) was decreased, cell death nevertheless occurred in the deficient cells. This was associated with abnormal mitosis.

In this chapter, we considered forms of cell death that are not apoptosis. Different paths to cell death are interesting (and may have fundamental repercussions for therapeutic intervention), but does it matter to the body how a cell dies? Consequences of cell death go beyond the cell, and, indeed, the mode of cell death can influence these sequelae. We explore such consequences next.

Chapter 9

The Burial
Clearance and Consequences

DEATH MATTERS

Cell death has consequences beyond simply the loss of the cell itself. Dying cells trigger their removal by other cells, and this can be accompanied by an inflammatory response. The cells can also induce other cells to proliferate to compensate for the loss in cell number. And, in vertebrates, they can signal to the immune system to regulate adaptive immune responses to foreign or altered proteins associated with the dying cell. Dying cells are not simply debris to be discarded; they are resources to be recycled, and this process of waste management is under tight controls.

In this chapter, we consider the ways in which dying cells are cleared from the body and the consequences for the organism after the cell is gone. As we will see, there are many possible outcomes, and these depend in part on the cell and how it dies.

DISPOSING OF THE CORPSE

In some cases, cells that die are simply sloughed off, and in vertebrates, this happens to the many cells that die in the skin and the gut epithelium as a consequence of normal tissue homeostasis. But most cells that die in the body have no place to go; they must be cleared by other cells.

Regardless of the mode of cell death, the corpses are engulfed[1] by phagocytic

---

[1] A quick note on the use of our term "engulfment" may be warranted. In general, when a cell eats a particle of any kind, this is referred to as *phagocytosis* (cellular eating). There are two main types of phagocytosis, *opsonization* and *macropinocytosis*. In opsonization, the phagocyte binds to the particle and effectively "zips" its membrane around it, stripping away anything loosely associated with it and excluding surrounding fluids. In contrast, macropinocytosis involves "gulping" the particle together with surrounding fluids and molecules. In the case of removal of cell corpses, both processes have been described (and have different consequences), leading to some confusion. For this reason, many who study the process prefer the term "engulfment" as a way to sidestep the issue. As we will see, there are several ways in which dying cells are removed by phagocytes.

cells, by which they are digested. Presumably, some breakdown products of the dead cell are recycled, but others represent a problem for the cell doing the clearing and may require a form of waste management. The phagocytes ("eating cells") are often professionals, such as macrophages and dendritic cells in vertebrates. However, they can also be amateurs, such as epithelial cells. Some of the consequences of cell death depend on which cell does the eating.

Although any dead cell can be removed by phagocytosis, apoptosis provides a means to clear the dying cell before plasma membrane integrity is lost, thereby avoiding secondary necrosis (see Chapter 8). As we will see, necrosis can promote inflammation through the release of intracellular contents. However, apoptosis involves effective packaging of the corpse and preprocessing of some of its components. The consequences of caspase activation—which leads to DNA fragmentation, compaction, and the blebbing off of pieces of the dying cell—can all be seen as preparation for clearance.

The signals produced by apoptotic cells echo Pete Townsend's "see me, feel me, touch me, heal me" from the classic rock opera *Tommy*.[2] The steps here, however, are "find me, bind me, eat me, clear me."

"FIND-ME" SIGNALS

Often, cells that die are simply engulfed and cleared by neighboring cells, and no recruitment of phagocytes is necessary. However, in the case of apoptosis, the facilitation of clearance can involve production of find-me signals by the dying cell that recruit macrophages and other phagocytes.

One such find-me signal is lysophosphatidic acid (LPA), which attracts macrophages, among its many other activities. Many cells that undergo apoptosis produce LPA in a caspase-dependent manner. Caspases cleave and activate calcium-independent phospholipase A_2 by removing an inhibitory domain on the enzyme, and this produces LPA (Fig. 9.1). There are also likely to be other, related (but currently unknown) mechanisms whereby caspases do this.

Dying cells can also release damage-associated molecular patterns (DAMPs, see Chapter 7). Some examples that act—probably indirectly—as find-me signals are ATP and another nucleotide, UTP, that is converted to UDP in the extracellular environment. Both of these bind to receptors on phagocytic cells. Another DAMP, released from necrotic cells, is uric acid, and this can also generate find-me signals. It is likely that in these cases, cells that are near the dying cell are activated by the DAMPs to produce molecules such as cytokines and chemokines, which are the actual signals

---

[2] Readers unfamiliar with the reference are urged to listen to The Who's classic rock opera *Tommy*, while reading this chapter. Others who haven't heard it in some time may wish to do the same.

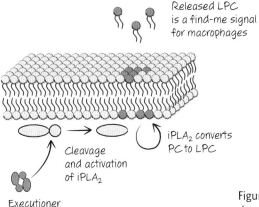

Figure 9.1. Executioner caspases induce a "find-me" signal.

that recruit more phagocytic cells. In general, the DAMPs that do this are released from necrotic cells, and whether they have roles in generating find-me signals for other forms of cell death has not been established.

"BIND-ME" SIGNALS ON DYING CELLS

As cells die, changes in the plasma membrane act as signals to engulf the corpse. The most important of these is a lipid, phosphatidylserine. In healthy cells, many plasma membrane lipids are arrayed asymmetrically, residing predominantly in either the inner or outer leaflet of the membrane. Phosphatidylserine is kept to the inner leaflet by an ATP-dependent aminophospholipid translocase that flips phosphatidylserine and other aminophospholipids from the outer to the inner membrane. If energy in the form of ATP is unavailable, or if the plasma membrane is disrupted, phosphatidylserine accumulates by diffusion in the outer leaflet. In cells undergoing apoptosis, however, externalization of phosphatidylserine occurs in a rapid, caspase-dependent manner, before plasma membrane integrity is lost. The scramblase responsible for this phospholipid scrambling has not been definitively identified and how caspases activate it is also unclear.[3]

Although phosphatidylserine is an important bind-me signal (probably the most important one for engulfment), other bind-me signals also appear on dying cells. Oxidation of lipids and changes in carbohydrates also occur and can contribute to recognition.

Calreticulin is a protein within the endoplasmic reticulum, and cells that die in different ways often expose this protein on the cell surface. This can also act as a bind-me signal for phagocytes. Calreticulin also appears to be important later on for subsequent responses to the corpse, as we will see.

---

[3] Several phospholipid scramblases have been identified, but at this point, their roles in apoptosis remain obscure.

DETECTING PHOSPHATIDYLSERINE WITH ANNEXIN V

The appearance of phosphatidylserine on apoptotic cells can be detected with a probe, annexin V, that binds to phosphatidylserine and can be coupled to fluorescent dyes for detection (Fig. 9.2). It is important to note, however, that annexin V is *not* how apoptotic cells are recognized by the body. Apoptotic cells can be stained with annexin V before they become permeable to other dyes, that is, before plasma membrane integrity is lost. In contrast, necrotic cells expose phosphatidylserine at the same time that they lose the plasma membrane barrier.

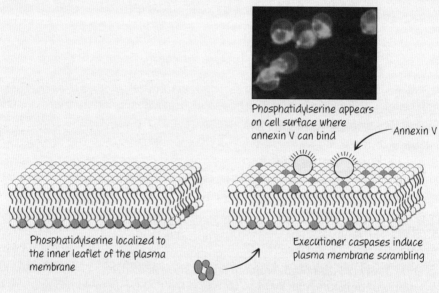

Figure 9.2. Plasma membrane scrambling and annexin V staining of apoptotic cells.

"DON'T-EAT-ME" SIGNALS ON CELLS

If phosphatidylserine exposure were the only signal to remove dying cells, then we would have a problem: Plasma membrane scrambling and externalization of phosphatidylserine occurs in perfectly healthy cells as well and may have roles in cell signaling. One way this occurs is through the activation of scramblases by calcium ions. Activation of lymphocytes, for example, induces a calcium flux that causes a transient exposure of phosphatidylserine.

But healthy cells (at least in vertebrates) appear to produce don't-eat-me signals that are rapidly lost from dying cells (Fig. 9.3). One don't-eat-me signal is CD47, a

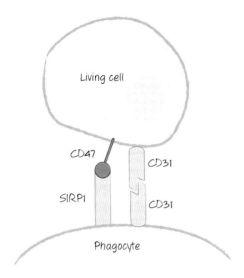

Figure 9.3. "Don't-eat-me" signals.

cell surface molecule expressed on all healthy cells. CD47 is recognized by a receptor, SIRP1α, on phagocytic cells, and this generates an inhibitory signal in the form of active tyrosine phosphatases. These appear to block phagocytosis as well as a range of other signaling pathways, and the CD47–SIRP1α interaction is also involved in other processes. During cell death, CD47 is rapidly lost from the cell surface or partitioned away from externalized phosphatidylserine.

Another don't-eat-me signal is the cell surface molecule CD31, which, when it interacts with CD31 on the phagocytic cell, generates an uncharacterized repulsion signal. The CD31 homotypic interaction is apparently disabled in dying cells.

BRIDGING MOLECULES RECOGNIZE "BIND-ME" SIGNALS

A number of soluble bridging molecules bind to dying cells because of the changes in the plasma membrane. Several of these bind to externalized phosphatidylserine on dying cells, and the most important of these is milk fat globulin-E8 (MFG-E8). Mice lacking MFG-E8, or animals injected with a dominant-negative mutant protein, accumulate apoptotic corpses. This is especially evident in one region of the lymphoid tissues, the germinal centers, where B lymphocytes proliferate during immune responses. In normal animals, germinal centers contain tingible body macrophages that turn out to be macrophages with engulfed apoptotic bodies that they are digesting (Fig. 9.4). Mice lacking MFG-E8 do not have these cells.

Several other bridging molecules bind to phosphatidylserine on dying cells. These include thrombospondin-1, Gas6, β1-GPI, and protein-S. Figure 9.5 illustrates the

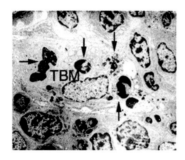

Figure 9.4. Tingible body macrophage (TBM) engulfing apoptotic cells (arrows).

array of bridging molecules that can bind to phosphatidylserine on dying cells.[4] Altered carbohydrates on the dying cells are bound by mannose-binding lectin (MBL) and the lung surfactant proteins A and D (SP-A and SP-D) (Fig. 9.5). It is likely that not all dying cells are bound by all of these bridging molecules, but clearly there is great redundancy in the system.

Two other bridging molecules may also be involved in the recognition of some dying cells. These are proteins in the complement pathway, a set of circulating proteins that function in immune responses to bacteria and other foreign invaders. C1q, which

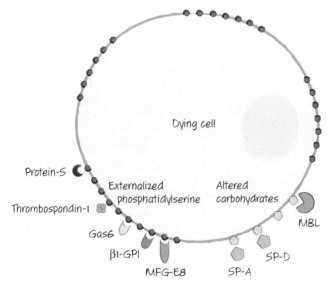

Figure 9.5. Bridge molecules.

---

[4] The bewildering variety of molecules and mechanisms for binding to dying cells can be taken as an indication of the importance of the process. It is also the consequence of an emerging field of research, where the importance of particular interactions is not fully understood. If the reader prefers, it is enough for now to let the message be "there are lots of bridging molecules" that can decorate dying cells.

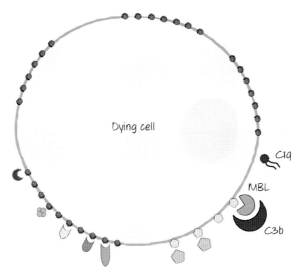

Figure 9.6. Complement components C1q and C3b can also act as bridge molecules.

binds to antibody clusters, appears to be involved, as is C3b, which normally binds directly to bacteria to trigger their removal (Fig. 9.6). In the case of dying cells, C3b appears to be bound and activated by MBL.

Of course, bridging molecules that decorate dying cells must themselves be recognized by receptors on phagocytic cells if they are to add engulfment. As we will see, there are also receptors that directly recognize the surface changes in the dying cells.

THE "TETHER AND TICKLE" MODEL

Why are there so many signals for uptake of dying cells? Relatively few of these actually trigger engulfment. A useful idea is that removal of cell corpses proceeds by a process of "tether and tickle." In this model, many of the receptors on phagocytes that recognize bridging molecules and the dying cells themselves simply tether the corpse. Other receptors, then, actually generate the responses leading to engulfment. These "tickle," that is, they produce intracellular signals that cause the phagocyte to eat the dying cell. This will be a useful idea as we continue our survey to include the receptors on the engulfing cells.

PHAGOCYTES HAVE RECEPTORS FOR "BIND-ME" SIGNALS AND BRIDGING MOLECULES

Many different receptors on the phagocyte bind to bridging molecules; others appear to bind directly to phosphatidylserine and signals on the dying cells. No one phagocytic cell has all of the relevant receptors, and therefore the rather complex picture that emerges may be a bit misleading. It is worthwhile to keep this in mind in the discussion that follows.

At least three receptors on phagocytes bind to phosphatidylserine and participate in the uptake of dying cells. These are TIM4 (T cell immunoglobulin- and mucin-domain-containing molecule 4—confusingly, on phagocytes), BAI1 (brain-specific angiogenesis inhibitor 1—confusingly, not brain specific), and stabilin-2 (Fig. 9.7). TIM4 appears to be particularly important; mice lacking TIM4 accumulate dying cells, and macrophages derived from bone-marrow precursors fail to take up apoptotic cells if they lack TIM4.

The phosphatidylserine-binding bridging molecule MFG-E8 binds to two integrins on the phagocyte surface: $\alpha_v\beta_3$ and $\alpha_v\beta_5$. Gas6 binds to a different receptor on the phagocyte, a receptor tyrosine kinase called MER. Both the integrins and MER signal, and therefore these probably "tickle" the cell to induce the subsequent engulfment. Macrophages from mice that lack these integrins or MER display at least partially defective binding to dying cells. TIM4 does not have a transmembrane region, and therefore *if* it "tickles" the phagocyte for uptake of corpses, how this happens is not clear.[5]

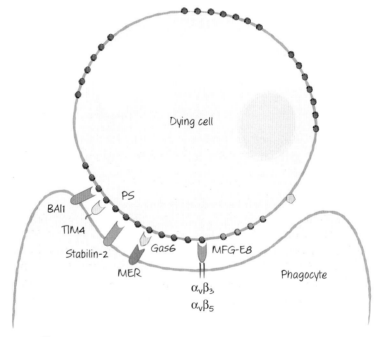

Figure 9.7. Recognition of phosphatidylserine (PS) by phagocytes.

---

[5] Another receptor for phosphatidylserine, PSR1, has also been described. However, subsequent studies have since shown that this is a nuclear factor and not directly involved in recognition of phosphatidylserine. It is mentioned here to help readers avoid unnecessary confusion when perusing the literature in this area.

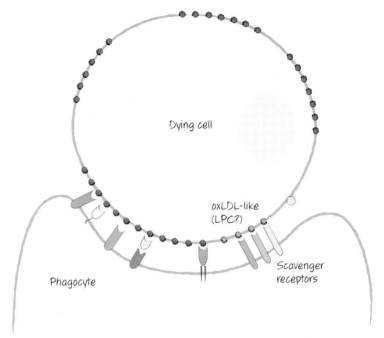

Figure 9.8. Scavenger receptors recognize dying cells.

Some of the other phagocyte receptors that are probably involved in tethering (although they might also have "tickle" functions about which we do not know) are the *scavenger receptors*, which include scavenger receptor-A, LOX1, CD68, and CD36. These normally bind to oxidized low-density lipids (oxLDLs) and clear them from the circulation. OxLDL-like sites on dying cells are also bound by these receptors. It is possible that these sites are actually lysophosphatidylserine (a find-me signal) that was not released from the dying cell (Fig. 9.8).

The complement molecule C3b is bound by complement receptors on the phagocyte. Because these clearly function in the engulfment of bacteria, it is likely that they similarly "tickle" phagocytes to engulf dying cells. In addition, C1q is also bound by a receptor composed of LRP1 (CD91) and the protein calreticulin (this time on the phagocyte). Calreticulin is also exposed on some dying cells, and it is likely that its interaction with LRP1 directly signals for engulfment. Figure 9.9 shows the array of receptors on phagocytes for dying cells and bridge molecules.

"BIND-ME" SIGNALS AND RECEPTORS IN OTHER ANIMALS

So far, we have focused entirely on the recognition of dying cells in vertebrate animals. However, much of what we know of the steps that follow binding has been gleaned

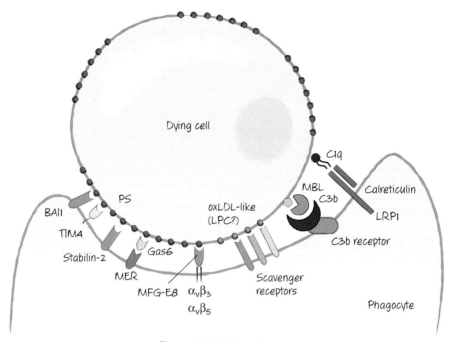

Figure 9.9. "Bind-me" receptors.

from genetic studies of *Caenorhabditis elegans*. These have uncovered two complementary pathways for recognition and engulfment of dying cells. Two of the genes identified (one for each pathway) may encode the actual recognition receptors, CED1 and CED7. CED1 is the nematode homolog of LRP1, the potential receptor for cell surface calreticulin, but we do not know whether calreticulin is the signal recognized by CED1.

CED7 is a cell surface molecule in the ATP-binding cassette (ABC) family. It is important in both the dying and engulfing cell and may engage in homotypic interactions (there are, of course, more complex possibilities). Some studies suggest that in mammals, two related proteins, ABC1 and ABC7, similarly function in engulfment of dying cells and contribute to phosphatidylserine externalization.[6] The recognition receptors of *C. elegans* are illustrated in Figure 9.10.

In *Drosophila*, an evocatively named protein, Croquemort ("undertaker"), can confer the ability to clear dying cells on a mammalian cell line on which it is otherwise lacking. Flies with mutations in the gene show defects in engulfment of apoptotic cells during development (Fig. 9.11). Croquemort is a homolog of the scavenger receptor CD36, which, as we have seen, is implicated in vertebrate recognition of dying cells.

---

[6] How ABC1 and ABC7 might function in either phosphatidylserine externalization or engulfment is not known, and therefore these were not included in the mammalian bind-me scheme discussed above.

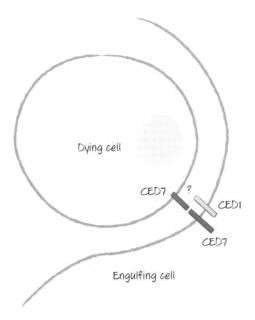

Figure 9.10. Recognition receptors CED1 and CED7 in nematodes. CED7 is important in both the dying and engulfing cells, whereas CED1 is important in the engulfing cell. It is not known with what either one binds on the dying cell. There is at least one additional receptor involved in recognition and engulfment of dying cells.

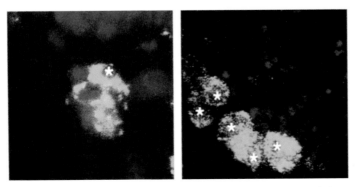

Figure 9.11. The *Drosophila* Croquemort mutant has defects in engulfment of dying cells. In the figure, phagocytic cells (green, with asterisks) take up small dying cells (red) in wild-type flies (*left*) but not in mutants in the gene (*right*).

"EAT-ME" SIGNALS PROMOTE ENGULFMENT

Engulfment requires extensive reorganization of the actin cytoskeleton to extend the phagocyte membrane around the corpse. The mechanisms that control this step are conserved in animals and involve small GTPases of the Rho family, especially RAC1 (CED10 in nematodes). This protein performs many functions in the cell, but we focus on its role in engulfment. In *C. elegans*, the two complementary pathways for engulfment converge on RAC1–CED10 (Fig. 9.12).

One of the pathways is linked to CED1 (LRP1 in vertebrates), which signals to RAC1–CED10 via the adapter molecule CED6, called GULP (get it?) in other organisms. Although we do not know whether LRP1 signals via GULP, another phosphatidylserine receptor, stabilin-2, appears to require GULP to induce engulfment (Fig. 9.13). How GULP–CED6 activates RAC1–CED10 is not known. In *Drosophila*, the CED1 homolog is called Draper, and this interacts with GULF–CED6 as well. But engulfment induced by Draper is also dependent on a tyrosine kinase,[7] and this

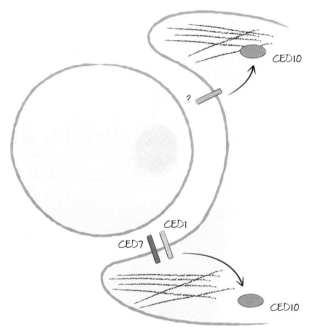

Figure 9.12. Two pathways converge on RAC1–CED10 to induce actin reorganization and phagocytosis. Although one pathway is linked to CED1 and CED7, the other pathway is initiated in nematodes by an unknown receptor.

---

[7] Intriguingly, this kinase is a homolog of the human kinase Syk, which is involved in the process of phagocytosis triggered by Fc receptors (the receptors on macrophages and other cells that bind antibodies). It is not known whether Syk participates in the uptake of dead cells in mammals.

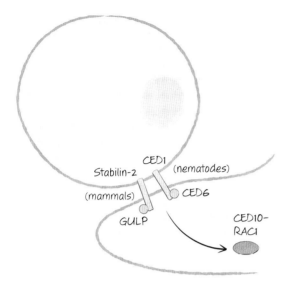

Figure 9.13. CED1 (nematodes) and stabilin-2 (mammals) engage an adapter molecule, CED6–GULP, following recognition of dying cells. CED6–GULP activates CED10–RAC1 by an unknown mechanism.

may in time provide a clue as to how this signaling complex functions.

The other pathway that converges on RAC1–CED10 is linked to an unknown receptor in nematodes, but to BAI1 in mammals. This signals via a complex of three nematode proteins: CED2, CED5, and CED12. CED2 is CRKII in other organisms. It has two SH2 (phosphotyrosine binding) domains and an SH3 (proline-rich binding) domain and performs numerous functions in regulation of actin. It interacts with CED5, called DOCK180 in other organisms, and CED12, called ELMO (for "engulfment and cell mobility") (Fig. 9.14).

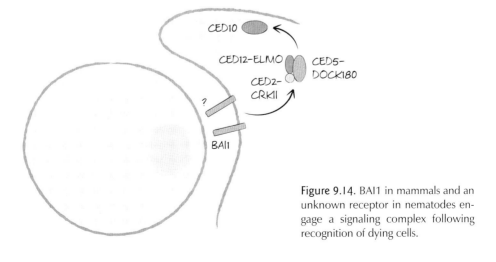

Figure 9.14. BAI1 in mammals and an unknown receptor in nematodes engage a signaling complex following recognition of dying cells.

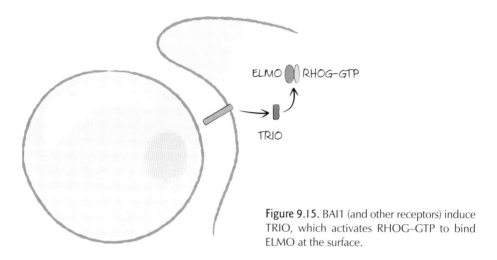

Figure 9.15. BAI1 (and other receptors) induce TRIO, which activates RHOG–GTP to bind ELMO at the surface.

Here is how it all appears to work.[8] DOCK180 is an usual type of guanine nucleotide exchange factor (GEF) thought to load GTP onto RAC1 by forming a scaffold that binds several components, including RAC1. DOC180 is normally inhibited by the binding of its SH3 domain to its RAC1-binding region. When an appropriate receptor, such as the phosphatidylserine-binding receptor BAI1, is activated, a protein called TRIO (another GEF) loads GTP onto the small GTPase RHOG. RHOG–GTP then recruits ELMO (Fig. 9.15).

ELMO then binds to the SH3 domain of DOCK180, allowing the latter to bind RAC1 as well as CRKII (Fig. 9.16). This complex, now localized to the contact site

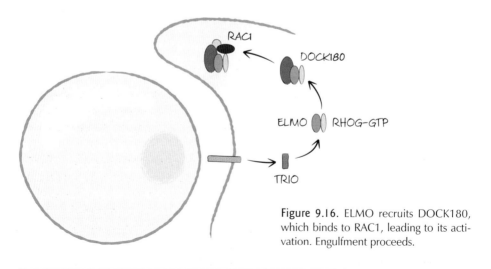

Figure 9.16. ELMO recruits DOCK180, which binds to RAC1, leading to its activation. Engulfment proceeds.

---

[8] This model is not quite complete. Our discussion assumes some understanding of GTPases and how they work; if this is obscure to the reader, it will not greatly matter for general understanding.

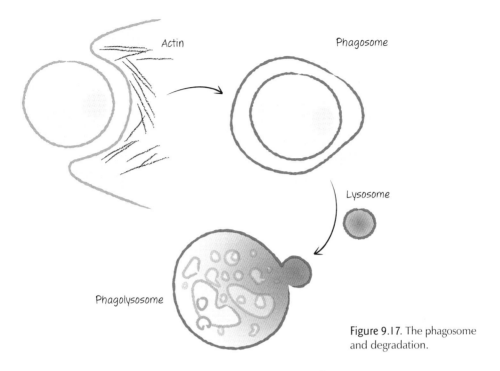

Figure 9.17. The phagosome and degradation.

where the phagocyte binds to the dying cell, activates RAC1. This, in turn, reorganizes the actin cytoskeleton.

The phagocyte membrane envelops the corpse to form a vesicle called the *phagosome*. The actin is reorganized, and the phagosome now carries its cargo off to lysosomes for disposal (Fig. 9.17).

THE "CLEAR-ME" PROCESS AND WASTE MANAGEMENT

Once the dying cell is engulfed in a phagosome, the engulfing cell must deal with the debris. This begins in the phagosome with the action of enzymes that degrade proteins and lipids, but the real work is performed when the phagosome fuses with lysosomes to form a *phagolysosome*.

Lysosomes contain an array of degradative enzymes including a nuclease, DNAse II. Apoptotic cells chop up their DNA by the caspase-dependent activation of the nuclease CAD (see Chapter 2). Cells lacking CAD or its inhibitor–chaperone iCAD do not effectively fragment their DNA. However, when such cells are engulfed by phagocytes, the DNA is nevertheless degraded. In contrast, mice lacking both CAD function and DNAse II have no DNA fragmentation of dying cells.

The iCAD–CAD complex is highly conserved in animals, but not present in nematodes. In the worm, a DNase in the phagosome digests the DNA. Digestion of DNA is an important waste management problem in the clearance of dying cells.

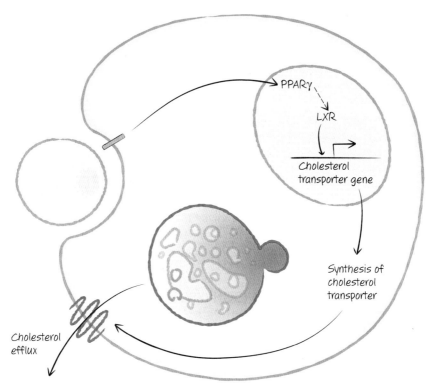

Figure 9.18. Uptake of dying cells induces cholesterol management. Activation of the PPARγ induces the expression of the nuclear receptor LXR, which in turn induces the expression of a cholesterol transporter.

Another waste management issue is cholesterol, the levels of which effectively double when a phagocyte takes up a dying cell. Recognition of phosphatidylserine on the corpse generates signals that activate PPARγ (peroxisome proliferator-activated receptor γ), an important regulator of cholesterol and other lipids in the cell. PPARγ shuts down a number of processes, including generation of cholesterol, and activates expression of a transporter to remove cholesterol from the cell. This is illustrated in Figure 9.18.

THE IMPACT OF DYING CELLS ON THE INNATE IMMUNE SYSTEM

Massive numbers of cells die in vertebrates all the time, and if we were to generate immune responses to these, we would face the severe consequences of autoimmunity.[9]

---

[9] To fully appreciate the impact of dying cells on the immune response, the reader will require a much more complete understanding of the immune system than can be provided here. Therefore, our overview is necessarily cursory.

But we must also be able to tell if the cell death is a consequence of infection, so that an appropriate response can be rallied.

The innate immune response induced by dying cells involves Toll-like receptors (TLRs) and Nod-like receptors (NLRs), the pattern-recognition receptors (PRRs) discussed in Chapter 7. When these are engaged by pathogen-associated molecular patterns (PAMPs) or damage-associated molecular patterns (DAMPs), they generate signals that include inflammasome activation as well as a variety of other effects. The latter include the production of cytokines and chemokines, molecules that control differentiation and functions of immune and inflammatory cells.

Cells that die by necrosis release a number of DAMPs that trigger inflammatory responses. These DAMPs include uric acid, ATP, and another DAMP that induces the production of cytokines but not inflammasome activation: high-mobility group protein B1 (HMGB1). HMGB1 interacts with RAGE (receptor for advanced glycation end products) and, perhaps, one of the TLRs, TLR4, on phagocytes to trigger such responses.[10]

In contrast, apoptotic cells seem to have a calming effect on these responses. Not only do they not have a chance to release their DAMPs before engulfment, in some settings they also actively suppress inflammatory responses. There seem to be different ways in which they do this. Some apoptotic cells release the cytokine transforming growth factor β (TGFβ), which inhibits many immune responses. Alternatively, some phagocytic cells (macrophages and dendritic cells) produce TGFβ following interaction with apoptotic cells.

Another way apoptotic cells inhibit inflammatory responses may be via the activation of PPARγ (discussed above). PPARγ inhibits the production of several inflammatory mediators, including leukotrienes and prostaglandins. Yet a third mechanism involves MER, the receptor for Gas6. MER also interacts with the type I interferon receptor, and this activates transcription of an intracellular inhibitory protein, SOCS1 (suppressor of cytokine signaling 1), which can interfere with signaling by TLRs. Figure 9.19 shows the different ways in which apoptotic cells can inhibit inflammation.

The extensive apoptosis that occurs during tissue turnover therefore does not evoke inflammatory responses. In contrast, what does is necrotic cell death, which occurs in response to damage or pathological conditions, and the inflammatory response is rallied to deal with any infectious agent that might as a consequence gain access to the body.

What happens with other forms of cell death? As we know, activation of the inflammasome and caspase-1 can lead to death in the cells that are stimulated, and this death is accompanied by the production of cytokines (interleukin-1 and interleukin-18) that can stimulate inflammatory responses. It is generally assumed that such cell death does not inhibit innate immune responses, but there are other more subtle possibilities. Finally, cells that undergo autophagic cell death or programmed necrosis are cleared from the body, but how these affect innate immune responses is not known.

---

[10] TLR4 is activated by bacterial lipopolysaccharides. Because these frequently contaminate proteins, there is always a concern that ligands that are identified for TLR4 are actually artifacts because of such contamination. Therefore, the ability of HMGB1 to directly stimulate TLR4 is controversial.

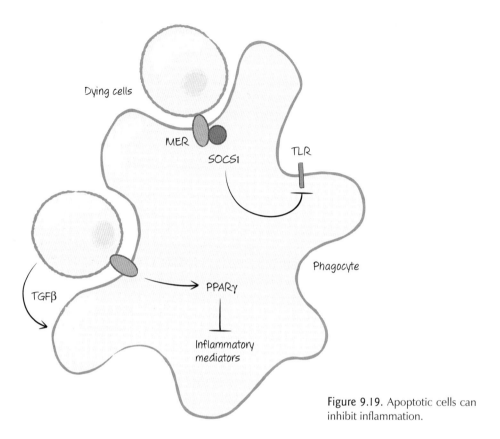

Figure 9.19. Apoptotic cells can inhibit inflammation.

CONSEQUENCES OF DYING CELLS FOR THE ADAPTIVE IMMUNE RESPONSE

When a phagocytic cell such as a macrophage or dendritic cell engulfs a particle, proteins associated with the particle are chopped into peptides and "presented" by the major histocompatibility complex (MHC) molecules on the cell surface. If a patrolling T lymphocyte recognizes the peptide, it can proliferate and generate an immune response, become refractory to stimulation, or actively inhibit immune responses. How it responds depends on other molecules on the presenting cell as well as the cytokines in the immediate environment.

PAMPs and DAMPs provide maturation signals to dendritic cells, the "professionals" that present peptides to T cells and determine the T cell response. Therefore, when cells that die by necrosis are eaten by dendritic cells, their proteins are processed to give peptides that are presented to T cells. Any peptides that are novel, say from an infectious organism that might have caused the necrosis, will be recognized, the T cells will proliferate, and an immune response will ensue.

Apoptotic cells may similarly generate immune responses but they can also actively inhibit them. The differences between *immunogenic* (the former) and *tolerogenic* (the latter) forms of apoptosis are not entirely clear. As with necrotic cells, the proteins associated with apoptotic cells are processed to give peptides that are presented on the surface of dendritic cells. Whether an immune response to any novel peptide occurs depends on several factors.

One such factor is the presence of calreticulin on the dying cell. Some agents that induce apoptosis promote the exposure of calreticulin, whereas others do not. The reasons for this are not known, but the exposure of calreticulin appears to correspond with immunogenic cell death.

Another factor is the DAMP HMGB1. HMGB1 is in all cells, and its release promotes immunity. However, this effect of HMGB1 can be blocked by oxidation (and perhaps other modifications) of the protein. During apoptosis, the caspase-dependent burst of reactive oxygen species can prevent the immunogenic effects, so that a tolerogenic response dominates. When they die, cells lacking HMGB1 tend to promote tolerance rather than an immune response, whether by apoptosis or necrosis (Fig. 9.20). In contrast, apoptotic cells in which the oxidation of HMGB1 is disrupted (e.g., by mutation of the caspase substrate NDUFS1, which is responsible for the reactive oxygen burst; see Chapter 2) tend to promote immune responses rather than tolerance.

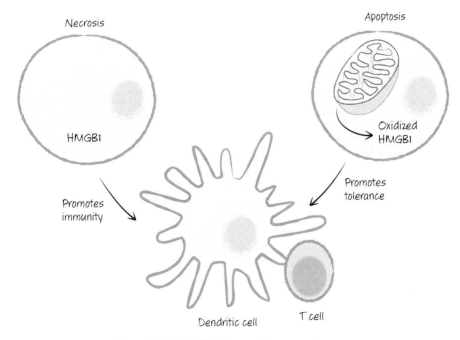

Figure 9.20. HMGB1 in immunity versus tolerance.

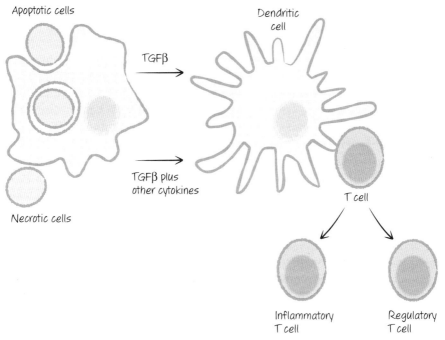

Figure 9.21. Dying cells influence T cell functional differentiation.

Cytokines also influence adaptive immune responses. As mentioned above, TGFβ is an inhibitory cytokine produced by some dying cells and by cells that engulf them. TGFβ induces T cells that encounter appropriate peptide antigens to differentiate into *regulatory T cells* (T_{reg}), which actively inhibit immune responses. However, other cytokines that can be induced by PAMPs can skew T cell differentiation toward a type of T cell that promotes rather than inhibits inflammatory responses (Fig. 9.21).

CLEARANCE AND DISEASE

Do all these complex decisions really matter? The short answer is yes. Mice with defects in apoptotic cell clearance, such as those lacking MER, MFG-E8, or C1q, not only accumulate extra cell corpses but also show a form of autoimmunity resembling human systemic lupus erythematosis (SLE), a disease in which the immune system produces antibodies against DNA and other nuclear components. In people with SLE, effective clearance of apoptotic cells frequently appears to be compromised. A diagnostic feature of SLE is the presence of so-called LE bodies in the circulation (see Fig. 9.22). These turn out to be free apoptotic corpses or corpses engulfed by macrophages (normally, clearance is sufficiently robust that such features are not seen in the circulation of healthy individuals).

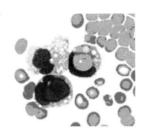

Figure 9.22. LE bodies. These are apoptotic cells in the circulation and are diagnostic of SLE.

The inhibition of immune responses by apoptotic cells may have therapeutic value. A routine procedure in tissue transplantation is to infuse cells from the donor individual into the recipient. It is likely that these cells undergo apoptosis and they may dampen immune rejection of the graft.

Another therapeutic benefit involves the use of apoptotic cells to treat autoimmune diseases. In animal models, apoptotic cells coupled to proteins that are targets of the immune cells causing such diseases have proven effective in preventing or ameliorating autoimmunity in animals. Alternatively, it is possible to mimic the signals generated by apoptotic cells: Drugs that activate a nuclear receptor downstream from PPARγ (see above) cure SLE in mice that have defective apoptosis. We will have to see if these approaches work in humans.

Finally, when we treat cancers, we often use therapies designed to trigger apoptosis in the tumor cells. It is possible that when such apoptosis is immunogenic, it contributes to eradication of the tumor.

COMPENSATORY PROLIFERATION

When cells die, we might expect this to generate a signal to make more of that cell type. Such *compensatory proliferation* has been verified in insects. Cells that undergo apoptosis in *Drosophila* produce factors that induce proliferation of surrounding cells. Interestingly, one of these factors, Decapentaplegic (Dpp), is a member of the TGFβ family (Fig. 9.23). This and the other factors[11] responsible for compensatory proliferation are produced in a manner that depends on the activation of caspases, but the precise mechanisms of release are not known.

In vertebrates, apoptosis induces compensatory proliferation of adult stem cells in several tissues. Although TGFβ is sometimes produced by dying mammalian cells, it does not appear to be important for this effect. Instead, caspase-mediated activation of phospholipase A_2 (see Fig. 9.1) leads to the generation of prostaglandin E_2 (PGE_2),

---

[11] The other pathways signal proliferation through Wingless (Wg; mammalian WNT is the homolog) and Hedgehog (Hh). Curiously, Dpp and Wg are produced in a manner that depends on the initiator caspase Dronc, but apparently not executioner caspases, whereas Hh signaling is dependent on the latter.

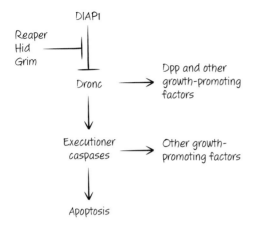

Figure 9.23. Caspase activation promotes compensatory proliferation in *Drosophila*.

which is responsible for the effect. Although caspases are required for production of this mediator from dying cells, necrotic cells can induce PGE_2 production by macrophages that engulf them,[12] and therefore it is likely that these, too, can induce compensatory proliferation of stem cells.

Compensatory proliferation in response to cell death can also occur less directly. Living cells rely on growth and survival factors, as we saw in Chapter 5. It turns out that our cells do not only rely on these but, instead, *compete* for them; cells often take up and actively degrade such factors. When a cell dies, there are more of such factors available for the living cells, and these can then promote proliferation until the normal number of cells is achieved.

This process can be readily shown by removing cells from some tissues, such as surgical resection of part of the liver, or by introducing small numbers of normal lymphocytes into an animal in which they are lacking. In either case, rapid proliferation of the healthy cells occurs until normal tissue levels are attained.

Because dying cells are not directly responsible for the proliferation of healthy cells, this effect is not generally called compensatory but is, instead, often referred to as *homeostatic proliferation*. As we will see in Chapter 11, even this indirect effect of cell death can have consequences for disease.

---

[12] However, as we noted, uptake of apoptotic cells seems to prevent production of prostaglandins. It is possible that in addition to limiting inflammation, apoptosis in mammals tends to limit compensatory proliferation, whereas necrosis promotes it.

Chapter 10

Cell Death in Development

DEATH CREATES FORM

Michelangelo once described his art like this: "For sculpture, I mean that made by removing; that by adding is similar to painting."[1] Although animals are not fashioned from solid blocks by the removal of unneeded parts, such a process does contribute to the generation of form and function in the embryo. Of course, development is an immensely complex process involving cell proliferation, differentiation, migration, and adhesion as well as cell–cell interactions and cell death. But here, we are concerned primarily with the latter.

The goal here is not to catalog all the ways that cell death contributes to animal development but, instead, to explore how specific cues can activate the core cell death machinery, to program cell death in development. That is, we examine how a cell can be *specified* for death.

As noted in Chapter 1, the term *programmed cell death* refers to cell death that occurs at a genetically prescribed point during development (and has since come to mean apoptosis in many investigators' lexicons) and removes extra cells that may be present for a variety of reasons. For example, they may have inductive or scaffolding functions in developing tissues or be involved in selection steps. We discuss examples of each. Alternatively, animals may undergo metamorphosis during their life cycles, and cell death is involved in the repatterning that occurs. The specification of cells for death by developmental cues is best understood in invertebrate systems, and we begin with some examples from these.

CELL DEATH IN NEMATODE DEVELOPMENT

During the development of nematodes, cells follow a strict program in which every cell produced, beginning with the fertilized egg, has a predetermined fate. 1090 cells

---

[1] In a letter to Benedetto Varchi, he wrote, "Io intendo scultura quella che si fa per forza di levare: quella che si fa per via di porre è simile alla pittura." Literally: "I mean, for sculpture, the one made by force of removing; the one through adding is similar to painting." (Thanks to Gerry Melino for the translation.)

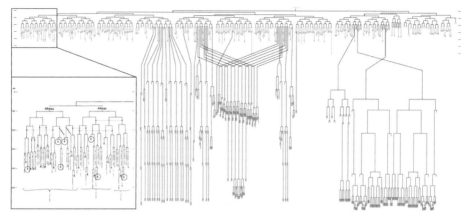

Figure 10.1. Developmental plan of the nematode *Caenorhabditis elegans*. During development, each cell that arises by cell division undergoes a set process of further division, differentiation, or cell death. (*Inset*) Part of the developmental plan, with cell deaths circled.

are produced and 113 of these die during embryogenesis. After this, another 18 cells die. Some of these cell deaths are indicated in the inset in Figure 10.1. In the adult hermaphrodite, about half of the developing oocytes die, but there are no other cell deaths. In males (the other sex), no cell death occurs in adults.

All of the cell deaths that occur are by apoptosis, except for one: the death of the so-called linker cell. This cell is involved in the development of the male gonad and appears to die by necrosis (Fig. 10.2). All of the other cell deaths depend on CED4 (the APAF1 homolog) and CED3 (the caspase).

Most of this cell death is in the neuronal lineages. The signals controlling the core apoptotic pathway are understood for two of these: the neurosecretory motor neuron (NMN) sister cells and the hermaphrodite-specific neurons (HSNs). The latter only die in developing male worms.

In NMN sister cells and the HSNs destined to die, the transcription factors HLH2 and HLH3 induce EGL1, the nematode BH3-only protein. Another factor, CES1, inhibits expression of EGL1 in the NMN cells that do not die. Yet another protein, CES2,

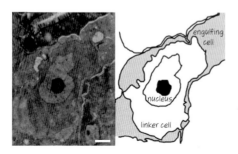

Figure 10.2. Linker cell death. Note that the dying linker cell has no features of apoptosis.

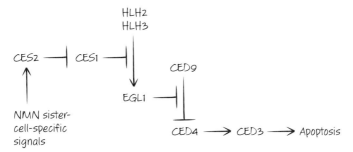

Figure 10.3. NMN sister cell death.

is expressed in the NMN sister cells specified for death, and CES2 transcriptionally represses CES1. As a result, the NMN sister cells die (see Fig. 10.3).

In males, the pathway controlling HSN death is similar. Again, HLH2 and HLH3 induce expression of EGL1, leading to cell death. In hermaphrodites, expression of EGL1 in the HSN cells is blocked by expression of another factor, TRA1, that is not expressed in males (Fig. 10.4).

Additional pathways exist in other cell lineages. These probably also act via regulators of the core apoptosis pathway, but how this occurs is less clear. In a cell called the tail-spike cell, the timing of cell death appears to be controlled by transcriptional regulation of the caspase CED3 (Fig. 10.5). Although CED4 is required for the developmental death of this cell, EGL1 and CED9 (which are upstream) appear to have only minor regulatory effects. The control mechanism is a transcription factor, PAL-1, that regulates CED3 and therefore apoptosis in this cell. EGL1 and CED9 therefore influence death but in this case are not targets of the specification.

The death of the tail-spike cell illustrates a concept that is useful for understanding the timing of cell death in development. If one or more components of the cell death pathway are limiting in a cell, factors that influence the expression and/or functions of these components can be critical for specification of the cell for death. Although only one cell in *Caenorhabditis elegans* appears to be specified this way, this form of cell death regulation is probably involved in developmental cell death in other organisms.

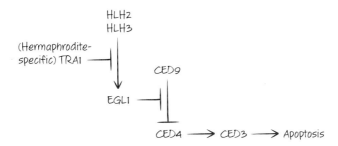

Figure 10.4. HSN cell death in males.

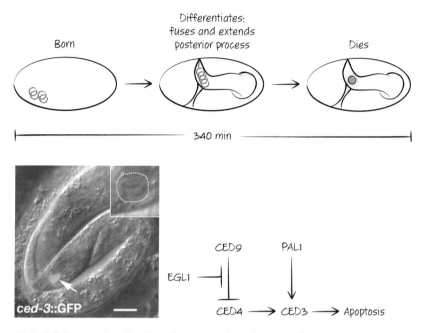

Figure 10.5. Cell death in the tail-spike cell. Death in this cell is controlled mainly by the transcription factor, PAL-1, that controls the expression of CED3, with little contribution from CED9 or EGL1. Image shows expression of CED3 in the dying cell.

CELL DEATH DURING METAMORPHOSIS IN *DROSOPHILA*

In insects, there is a dramatic change in morphology when the larva pupates and undergoes metamorphosis to form an adult. In *Drosophila*, this is triggered by two waves of the steroid hormone ecdysone, whose receptor is a ligand-dependent transcription factor. The second wave induces the expression of the transcription factors BR-C, E74, and E93. These up-regulate the expression of several genes involved in cell death, including those encoding ARK, Dronc, and Drice (the fly APAF1 homolog, caspase-9 homolog, and executioner caspase, respectively). These are probably not sufficient to cause cell death in metamorphosis, because the caspases are inhibited by DIAP (see Chapter 3). However, the transcription factors also induce the expression of two DIAP inhibitors, Reaper and Hid, and as a consequence, apoptosis is engaged (Fig. 10.6).

Cell death by apoptosis proceeds in the anterior muscles, larval midgut, and larval salivary glands. Apoptosis is not responsible for all (or even most) of these cell deaths. Much of the cell death that occurs during insect metamorphosis displays characteristics of autophagic cell death (see Chapter 8). Figure 10.7 shows autophagic cell death in the metamorphosing salivary gland.

CELL DEATH IN DEVELOPMENT   153

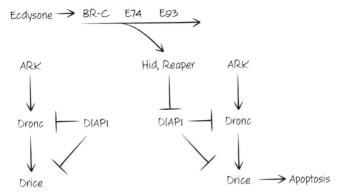

Figure 10.6. Activation of the apoptosis pathway by ecdysone.

It turns out that BR-C, E74, and E93 also induce the expression of several proteins in the autophagy pathway, including ATG5 and ATG7. As we discussed in Chapter 8, autophagic (type II) cell death is accompanied by autophagy but is often not dependent on this process. In this case, however, the autophagy pathway is required for the cell death that occurs, and flies with defects in autophagy display less cell death in the salivary glands during metamorphosis (Fig. 10.8). We do not know precisely how the autophagy pathway kills these cells, but several tissues show this effect.

In addition to expression of proteins that promote death by apoptosis and autophagy, ecdysone triggers expression of Croquemort, the *Drosophila* CD36 homolog involved in phagocytosis and clearance of dying cells (see Chapter 9). Presumably, these are expressed in cells that do not die, so that they can participate in corpse removal during metamorphosis. There is another possibility, however: It may be that autophagy and phagocytosis in the dying cells cooperate in their death. Flies with defects in phagocytosis accumulate living cells in the metamorphosing salivary gland, and intriguingly, the CED1 homolog Draper is required for autophagic cell death in this tissue.

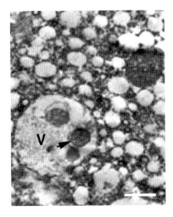

Figure 10.7. Type II death in the *Drosophila* salivary gland. Autophagosomes in dying cells are abundant.

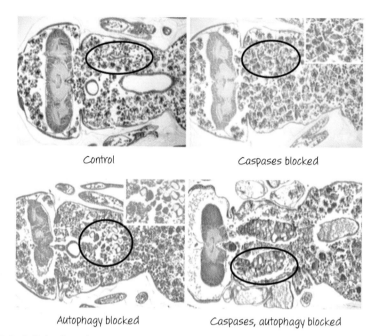

Figure 10.8. Cell death in the metamorphosing salivary gland. Inhibition of either caspases or autophagy promotes cell survival (circled areas); inhibition of both produces additive effects.

CELL DEATH IN VERTEBRATE DEVELOPMENT

Cell death occurs in many cell types throughout vertebrate development, and in the majority of cases, this is by apoptosis. In invertebrates, cell death has roles in formation of structures, removal of structures, and control of cell number. In vertebrates, it can also have fundamental roles in selecting cells for emergent functions of a cell population.

A classic example of cell death in the formation of a structure is the generation of digits in the vertebrate limb. During development, a limb paddle (or *autopod*) forms in which tissue fills the areas between the skeletal elements. Bone morphogenic proteins (BMPs), members of the TGFβ family, trigger an apoptotic pathway in the interdigital webs, but how this specifies cells for death is not known. Figure 10.9 shows examples from chick and duck development in which the extent of apoptosis creates characteristically different outcomes.

Although the precise mechanisms have not been delineated, this interdigital cell death is almost certainly mediated by the mitochondrial pathway of apoptosis. Mice lacking the pro-apoptotic BCL-2 family effectors BAX and BAK usually die as embryos, but rare mice that survive retain persistently webbed feet (Fig. 10.10). Remarkably, this is also seen in mice that lack two of the BH3-only proteins, BIM and BMF.

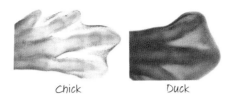

Figure 10.9. Cell death in developing chick and duck autopods. Dying cells stained with red dye in the chick interdigital webs are seen only at the web tips in ducklings. The apparent red staining on the right is an artifact; only the tips of the webbing contain dying cells.

Mice lacking APAF1 or caspase-9 show a delay in digit formation during embryogenesis, but the cells nevertheless die by caspase-independent cell death (see Chapters 4 and 8).

Another example of cell death creating form is cavitation, the process by which some areas hollow out during development. Examples include the formation of the pro-amniotic cavity just after implantation in mammals and the hollowing of ducts in the mammalian breast. An unknown prodeath signal is generated, but cells that are in contact with the basement membrane receive survival signals that block the death. Only the cells in the center undergo apoptosis (Fig. 10.11).

In the case of the mammalian breast, the cell death signal appears to be provided by the BH3-only proteins BIM and BMF. Anti-apoptotic effects of MCL-1 may prevent the death of cells in contact with the basement membrane.

In cavitation, cells that move away from the basement membrane and its extracellular matrix may undergo anoikis (see Chapter 5) because they no longer receive protective survival signals. The concept is a useful one and probably applies to many other developmental events. The survival signals provided by growth factors and the extracellular matrix help to define the boundaries to which a tissue can extend.

Other interesting examples of developmental cell death in vertebrates are the loss of the tail and restructuring of the intestine in amphibian metamorphosis (Fig. 10.12). Here, the signal for metamorphosis and cell death is thyroid hormone. This

With BAX and/or BAK

Without BAX or BAK

With BIM and/or BMF

Without BIM or BMF

Figure 10.10. Webbed feet in mice lacking BAX and BAK or BIM and BMF.

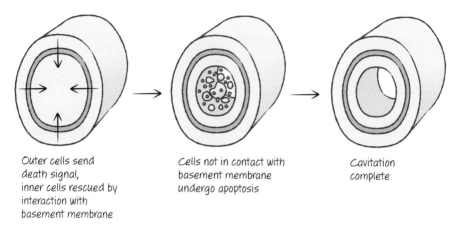

Outer cells send death signal, inner cells rescued by interaction with basement membrane

Cells not in contact with basement membrane undergo apoptosis

Cavitation complete

Figure 10.11. Basic cavitation.

triggers a gene expression program in which several caspase genes are up-regulated, and there are changes in the expression of BCL-2 family proteins as well as a variety of tissue proteases. How cells are specified for death is not well understood, nor is the initiation of apoptosis. It is likely that the mitochondrial pathway of apoptosis is important, because removing the amphibian caspase-9 homolog severely perturbs metamorphosis.

APOPTOSIS AND SELECTION IN DEVELOPMENT

Collections of cells can have emergent properties that far exceed the sum of their parts. By overproducing cells and then selecting for functions, the body can take advantage of such emergent properties. Selection can be positive, negative, or both, and most (if not all) selection involves cell death. Two examples of such selection in vertebrate development produce connectivity among neurons and self–nonself discrim-

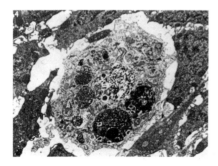

Figure 10.12. Frog metamorphosis and cell death. An apoptotic cell in the tail of a metamorphosing tadpole.

ination in the immune system. The latter probably represents the best example available of specification for cell death in a vertebrate system.

SELECTING NEURONS

In the developing nervous system, the axons of neurons extend to form connections with the dendrites of other neurons, and proper connectivity is essential for higher-order functions. One important way this may come about is proposed in the *neurotrophism model*. In this model, more neurons than will be ultimately used are produced, and these extend their axons toward another "target" population of neurons. Those that hit their targets then receive signals necessary to sustain their survival, whereas those that do not hit their targets die (Fig. 10.13).

Neurotrophic survival factors, such as nerve growth factor released by the target cells, bind to receptors (e.g., TrkA) on the recipient axon and are transported to the cell body. There, the signals generated by the receptor are easily transmitted to the mitochondria and the nucleus. We can mimic this effect by culturing neurons and then depriving them of survival signals. When developing sympathetic neurons are deprived of nerve growth factor, they undergo apoptosis. This apoptosis depends, in part, on the BH3-only proteins BIM, BAD, and HRK and the pro-apoptotic effectors

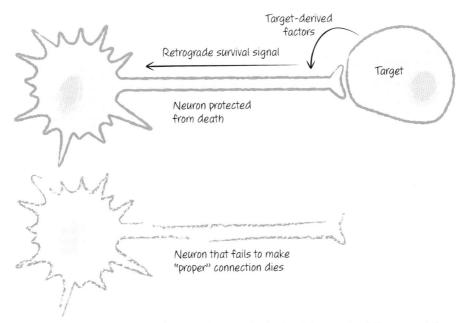

Figure 10.13. Cell death as a mechanism of neuronal selection. Neurons that fail to contact the appropriate target cell do not receive factors that promote their survival.

BAX and BAK (see Chapter 5). MOMP then occurs, and caspases are activated via the mitochondrial pathway. As discussed in Chapter 5, MOMP can result in caspase-independent cell death in many cell types. However, neurons deprived of nerve growth factor can survive post-MOMP if the mitochondrial pathway is blocked or disrupted. This may be why mice lacking APAF1, caspase-9, or caspase-3 sometimes have brain abnormalities and excess neurons (see Chapter 8). Intriguingly, this phenotype is dependent on the genetic background of the mouse. In some strains, mice lacking these components of the mitochondrial pathway do not show severe developmental defects. At this point, we do not know the reason.

There is a problem with the view that the mitochondrial pathway of apoptosis is crucial for neuronal development. Mice lacking BAX and BAK cannot engage the mitochondrial pathway, but they sometimes survive, and although these animals display increased numbers of neurons, their behavior is not obviously disturbed. We have no good explanation for this. Either cell death is not absolutely essential for neurotrophic selection or there are other ways in which these cells die.

APOPTOSIS IN LYMPHOCYTE SELECTION

The immune system has a remarkable ability to discriminate between self tissues and foreign invaders, and this self–nonself recognition is essential for survival in an opportunistic world.[2] How this comes about involves a number of mechanisms, but we will focus on a central process that specifies developing lymphocytes to undergo apoptosis if they have the potential to recognize self tissues that would obviously endanger the organism.

Vertebrate T lymphocytes express specialized T cell receptors capable of recognizing small peptides presented to them on the surfaces of other cells (see Chapter 9). These receptors are generated in developing T cells by a random process involving rearrangements of gene segments that provide each T cell and its progeny with copies of a unique T cell receptor of indeterminate specificity. The immature T cell progenitors arise throughout life from hematopoietic stem cells that move to the thymus and differentiate into T cells.

At one point in the process, after the T cell receptor has been generated and expressed on the cell surface, the cell undergoes a test in the form of the ligand for its receptor: a peptide bound to an MHC molecule on another cell (see Chapter 9). If

---

[2] The concept of self–nonself discrimination by the immune system was proposed at the beginning of the 20th century and was so persuasive an idea that almost 50 years passed before we recognized that some diseases are actually caused by a failure of the mechanisms involved, leading to autoimmunity. It was around this time that we first realized that it is only an approximation of self–nonself discrimination that is learned by the immune system. It is a very complex process, and we concern ourselves here with only one of the steps in the learning mechanism and how cells are specified for apoptosis. Interested readers would do well to consult an immunology textbook for a detailed treatment.

the T cell recognizes this ligand, it undergoes apoptosis and is removed from the population. This process is called negative selection. In effect, the ligand for the T cell receptor on a given cell can specify the cell for death at this stage of lymphocyte development. Later, after the cell matures, recognition of ligand by the T cell receptor instead produces an immune response (Fig. 10.14).

Negative selection helps self–nonself discrimination emerge in T cells. Peptides derived from proteins from the organism itself are always present, and therefore negative selection ensures that no T cell matures that can make immune responses to these self-peptides. Novel peptides that appear (e.g., during infection) are detected by T cells that developed when these were not present. Therefore, superficially, the discrimination is between peptides that are always present (self) and those that are only sometimes present (nonself).

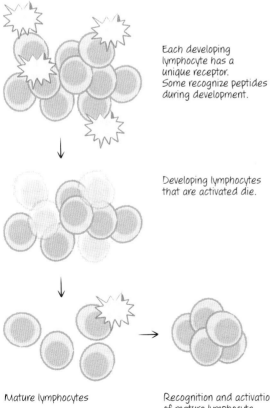

Each developing lymphocyte has a unique receptor. Some recognize peptides during development.

Developing lymphocytes that are activated die.

Mature lymphocytes

Recognition and activation of mature lymphocyte induces proliferation, immune response, and memory.

Figure 10.14. The concept of negative selection. Cell death of activated lymphocytes during their development produces emergent self–nonself discrimination.

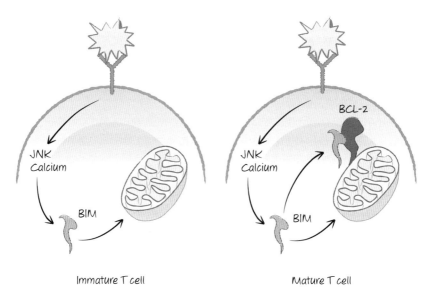

Figure 10.15. BIM and negative selection in T cells. T cell activation induces JNK and elevated calcium, both of which stabilize BIM. In immature T cells, BIM induces apoptosis, but in mature T cells, this is blocked by BCL-2.

How T cell receptor signaling results in apoptosis is at least partially understood. When stimulated, the receptor complex generates signals that cause a rapid elevation of calcium ions in the cytosol. Calcium stabilizes the BH3-only protein BIM, allowing it to accumulate.[3] In addition, signaling also activates the kinase JNK, which further stabilizes BIM and also phosphorylates BCL-2, and this decreases the anti-apoptotic activity of the latter (BIM is probably also transcriptionally induced in these cells, but the mechanisms have not been delineated). BIM then triggers the mitochondrial pathway of apoptosis. Mice that lack JNK or BIM do not display apoptosis associated with negative selection. However, the surviving cells do not mature to produce autoimmune disease; so clearly, there are additional checkpoints that control this important process if apoptosis fails.[4] T cell negative selection in the thymus nevertheless represents a clear example of specification for cell death mediated by a well-defined selection signal (Fig. 10.15).

Once the T cell matures, BIM remains in place, but now, elevated levels of the anti-apoptotic protein BCL-2 appear to keep the cell alive. Mice that lack BCL-2 de-

---

[3] In Chapter 5, we described some ways in which BIM is regulated. Another way is through calcium, which appears to activate a phosphatase to remove a phosphate group from BIM that otherwise promotes its degradation.

[4] In some experimental systems, defects in the recognition system that initiates negative selection permit such cells to mature, and the consequences include devastating multiorgan autoimmunity. It is likely that in the absence of BIM-mediated apoptosis, stimulated cells differentiate into immunosuppressive T_{reg} cells that inhibit any autoreactivity (see Chapter 9).

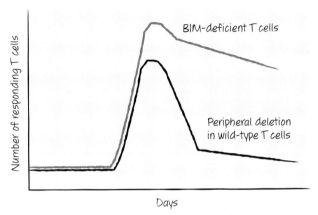

Figure 10.16. Peripheral deletion and BIM.

velop T cells, but these die in the first weeks after birth. If these animals also lack even one allele of BIM, however, the T cells mature normally and survive to function in immune responses.[5] In normal individuals, the recognition of a ligand by a T cell causes it to proliferate rapidly, but this is followed a few days later by a population contraction phase because of apoptosis of the clone, and this limits the immune response. This contraction phase is also because of BIM, and animals lacking BIM do not show this effect (Fig. 10.16).

Those cells that persist mediate immune memory to guard against the appearance of the ligand again. We do not know what factors determine the cells that die after the initial expansion versus those that persist to preserve immune memory.

Our foray into cell death in development has touched on how apoptosis and other forms of cell death are specified and how they function in some normal organismal processes. But cell death and survival also figure prominently in disease. We have seen how cell death is involved in ischemic injury and other forms of damage (see Chapter 8). We now turn to a discussion of how cell death and its regulation contribute to another important problem of defective cell and tissue homeostasis, cancer.

---

[5] The story is more interesting than related here, with additional insights into apoptosis and development. Mice that lack BCL-2 develop fairly normally to birth, but turn gray within three weeks and die of kidney failure a few weeks later. If the animals lack one allele of BIM, they survive (but have gray hair). If they lack both alleles of BIM, their coat color is normal.

Chapter 11

Cell Death and Cancer

WHY IS CANCER RARE?

Most of us have been touched by cancer, whether directly or in people close to us; so at first, this question seems bizarre. But humans are composed of about 10^{14} cells, most of them turning over continuously, and given that mutations arise at a frequency of about 1 in a million (and note that we produce about a million cells every *second*), why are we not all riddled with cancers every day?

The answer, of course, is that multiple checkpoints are in place to prevent cancers from occurring. In fact, if we try to produce cancer experimentally—for example, by introducing a known cancer-causing gene (an oncogene) into a tissue—often cancers arise only after a delayed period and from a single cell that has acquired additional mutations. These checkpoints are efficient.

Cancer is one of two diseases in which a single, dysregulated cell can, in theory, lead to lethal disease (the other is autoimmune disease caused by one lymphocyte and its progeny). It is better, in either case, to sacrifice the cell than try to manage the consequences. The induction of apoptosis is therefore a key strategy for preventing the emergence of cancer.

APOPTOSIS AND PROLIFERATION

It is a frequent misconception that if the cell cycle is suddenly halted, this directly activates apoptosis. However, any links between the machinery of the cell cycle and those of the core apoptotic pathways are at best obscure (one such link is mitotic catastrophe; see Chapter 8). The cell cycle, per se, does not directly regulate apoptosis. That said, there are connections between apoptosis and the signals that *engage* the cell cycle, and these are fundamentally important for our discussion of cancer.

Myc is a nuclear factor required for entry into, and probably passage through, the cell cycle in animal cells. Mammals have three forms of Myc (Myc, L-Myc, and N-Myc) and all function as transcription factors that cause cells to proliferate. It is there-

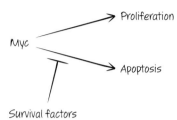

Figure 11.1. Antagonistic pleiotropy.

fore logical that if Myc is expressed, cells will accumulate. But things are not that simple. The expression of Myc can also induce cells to undergo apoptosis.

Relatively low levels of Myc are sufficient to engage the cell cycle, but the apoptosis that would also be triggered by Myc in these cells is blocked by survival signals. Any of a variety of extracellular molecules that engage cell surface receptors can produce anti-apoptotic signals (Fig. 11.1). For example, activation of AKT by survival factors[1] induces the expression and/or stabilization of anti-apoptotic BCL-2 proteins and the repression and/or destabilization of pro-apoptotic BCL-2 proteins (see Chapter 5).

The underlying idea is termed *antagonistic pleiotropy*.[2] If a cell activates the cell cycle (e.g., by engaging Myc), it will die unless *other* cells provide signals instructing it to survive and thereby, generate increasing numbers of cells as it proliferates. Therefore, the "society of cells" collectively determines how much expansion is allowed, based on the availability of survival factors.

Is it simply apoptosis that restricts cell accumulation? One approach to answering this is to replace survival factors (that have other effects as well, including metabolic effects) with an anti-apoptotic BCL-2 protein. This can also promote cell accumulation in this setting by blocking cell death (Fig. 11.2).

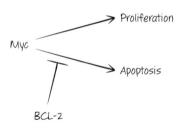

Figure 11.2. Antagonistic pleiotropy is regulated by BCL-2 proteins.

---

[1] Survival factors include cytokines (such as interleukin-2), growth factors (such as epidermal growth factor), and hormones (such as insulin) that bind to receptors on the cell surface and induce signals that preserve cell survival.

[2] There is another form of antagonistic pleiotropy that is an important model for the evolution of aging, and the two may be related. This other form does not concern us in this discussion, but it is worth looking up.

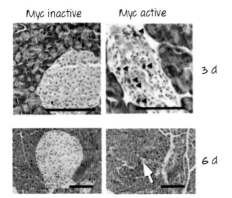

Figure 11.3. Myc in pancretic islets causes apoptosis. Myc was activated in pancreatic islets and apoptosis was assessed: (*Top right*, apoptotic cells marked with arrowheads). After 6 d, islets had involuted (*lower right*, white arrow).

The apoptosis-inducing effects have been dramatically shown in mice in which an inducible form of Myc was expressed in the β-islet cells of the pancreas (the cells that make insulin). Following activation of Myc, the cells died by apoptosis and the animals became diabetic (Fig. 11.3).

However, if the cells were also provided with an anti-apoptotic signal such as BCL-2, the effect of activating Myc was very different. In this case, every islet expanded dramatically and displayed all of the features of a tumor (Fig. 11.4).

This extremely informative model tells us that two signals (one for proliferation and one to block apoptosis) may be all that is needed to promote cancer. Tumors, however, are not generally this simple, and many other changes contribute to a tumor's aggressiveness and ultimately harm. But the underlying concept is a useful one. Blocking apoptosis can have profound effects that promote oncogenesis.

One type of cancer appears to be caused predominantly by this anti-apoptotic effect. As B lymphocytes develop, rearrangements in the DNA occur that are essential for the generation of antibody diversity. Sometimes, the rearrangement goes awry,

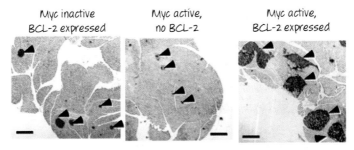

Figure 11.4. The activation of Myc in pancreatic islets expressing BCL-2 results in tumor-like expansion. Islets (arrowheads) expressing BCL-2 appear to be normal (*left*). Activation of Myc without BCL-2 for 7 d causes islet involution (*middle*). Activation of Myc with BCL-2 results in pronounced tumor-like expansion (*right*).

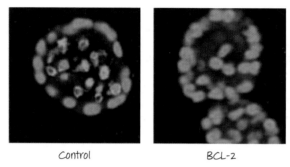

Control　　　　　　　BCL-2

Figure 11.5. Breast epithelial cells resistant to apoptosis fill the acinus. Control cells undergo apoptosis (green), whereas those expressing BCL-2 do not.

leading to constitutive expression of BCL-2. This can result in a cancer called follicular lymphoma. It appears that these cells fail to die, and they increase in number when they receive occasional signals to proliferate.[3]

Inhibition of apoptosis allows cells to not only proliferate but also to invade areas where normal cells would die. Adherent cells undergo anoikis if they lose contact with the basement membrane (see Chapter 5). However, cells that are resistant to apoptosis have no such requirement. An example can be found in the acinus of the breast. The center (lumen) is devoid of cells, because any cell that enters the lumen dies by anoikis. Nontransformed breast epithelial cells that express anti-apoptotic BCL-2 proteins, however, can fill the acinus (Fig. 11.5).

TUMOR-SUPPRESSOR PROTEINS PREVENT UNRESTRICTED PROLIFERATION

Signals that drive cells into the cell cycle are offset by tumor-suppressor proteins that prevent uncontrolled proliferation. Most tumor suppressors that have been identified are involved in signaling pathways affecting apoptosis, cell cycle, DNA repair, or other processes necessary for cell survival, proliferation, and the maintenance of the genome.[4]

One tumor-suppressor protein appears to dominate over all others in terms of its importance—the transcription factor p53. About half of all known tumors have mutations in the gene that encodes p53 (*TP53*), and those that do not have mutations in other components of the pathway responsible for its function. As we will see below, p53 has a number of functions, including promotion of apoptosis.

---

[3] BCL-2 (B cell lymphoma-2) is named for its presence at the breakpoint of some genetic translocations in B cell lymphomas. It is unlikely that the constitutive expression of BCL-2 is the only factor contributing to follicular lymphoma, and the example is therefore almost certainly an oversimplification.

[4] Generally, for a protein to be categorized as a tumor suppressor, it is insufficient to demonstrate that its expression can prevent oncogenesis; its elimination must be shown to promote cancer in some setting.

ACTIVATION OF p53

p53 is a transcription factor that induces the expression of a wide variety of genes, including genes involved in apoptosis. One target of p53 encodes a protein called mouse double minute 2 (MDM2). MDM2 is a ubiquitin ligase that binds p53 and modifies it for proteasomal degradation. Usually, p53 induces just enough MDM2 to ensure that neither is found at any detectable level in normal cells (Fig. 11.6).

Figure 11.6. The p53-MDM2 feedback loop.

Anything that disrupts the interaction of MDM2 with p53 causes the accumulation of p53. The disruption of the inhibitory interaction between MDM2 and p53 occurs in two fundamentally different ways. One involves a competing protein and the other involves modification of p53 itself.

Oncogenes capable of driving a cell into an unregulated cell cycle induce the expression of a protein called ARF.[5] ARF binds to MDM2 and thereby blocks its interaction with p53, allowing the latter to accumulate and perform its functions (Fig. 11.7). Tumors that have functional p53 very often have mutations in or otherwise silence the *ARF* locus. Thus, even if p53 is present, the absence of ARF prevents its accumulation in response to the oncogenes.

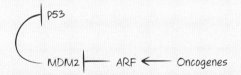

Figure 11.7. Oncogenes induce p53.

ARF also controls another protein, another E3-ubiquitin ligase that targets p53 for degradation. Thus, ARF stabilizes p53 by not only blocking its interaction with MDM2 but also by blocking the function of this second ligase. Remarkably, this protein, ARF-BP1, turns out to be MULE,[6] one of the E3 ligases that target the anti-apoptotic BCL-2 protein MCL-1 for degradation (see Chapter 5). Therefore, the ARF-BP1–MULE E3-ubiquitin ligase seems to have opposing functions: promoting apoptosis by causing the degradation of MCL-1 and inhibiting apoptosis by promoting p53 degradation. How this fits into a broader scheme of cell regulation is not obvious, and clearly there are more complex interactions around this cross talk between apoptosis and p53 control.

---

[5] ARF is an unusual protein in that its coding sequence overlaps that of another protein, INK4A, in a different reading frame (in fact, ARF stands for alternative reading frame). Interestingly, INK4A is a tumor suppressor that blocks the cell cycle.

[6] The proteins were identified and named independently but turned out to be the same protein.

(Continued)

(Continued from previous page)

When active, p53 controls the expression of other proteins that regulate MULE activity, including translationally controlled tumor protein (TCTP), which is repressed by p53. Among its many functions, TCTP binds to MCL-1 and prevents its degradation by MULE. Therefore, when p53 is activated, the levels of TCTP decrease, leading to less MCL-1. Because MCL-1 is anti-apoptotic (see Chapter 5), this interaction can promote apoptosis. As we will see, the role of p53 in apoptosis is considerably more complicated than this.

DNA DAMAGE INDUCES p53

Damage to DNA is a potent trigger for events that include DNA repair, cell cycle arrest, and/or apoptosis. Activation of p53 is an important component of this response, but this generally occurs independently of ARF. Instead, a second way in which p53 is activated is by phosphorylation of its amino-terminal region, to which MDM2 binds. Two related kinases, ATM and ATR, respond to damaged DNA either by directly phosphorylating p53 or by activating kinases Chk1 and Chk2 that phosphorylate p53 on a nearby site. These phosphorylation events interfere with the binding of MDM2 to p53, and the latter accumulates (Fig. 11.8).

Many other kinases (and phosphatases) target p53, and in most cases the effects of the phosphorylation are not known, but this may affect which genes that p53 can activate. p53 is also modified in other ways, including by acetylation and sumoylation, and these probably influence its transcriptional activities.

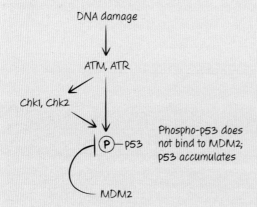

Figure 11.8. DNA damage induces p53.

INDIVIDUALS WITH DEFECTIVE p53 ARE PRONE TO CANCER

People with Li-Fraumeni syndrome[7] carry a defective p53 allele (two defective alleles is unusual and generally embryonically lethal in humans) and such individuals are extremely prone to cancer, particularly breast cancer, sarcomas (soft tissue and bone),

---

[7] Named for Frederick Pei Li and Joseph F. Fraumeni, who identified this syndrome.

brain tumors, and acute leukemias. In the cancers that arise, the other p53 allele, which was originally wild type, has mutated, so the cells no longer have functional p53. It is not unusual for such individuals to have two or more different cancers in their lifetimes.

Similarly, mice lacking both alleles of p53 always develop tumors. These tend to be tumors of thymic origin, but others also arise. Specific deletion of p53 in particular tissues can also promote cancer, especially if DNA-damaging agents (such as γ radiation) are used to promote mutagenesis.

Lack of even one allele of p53 in mice usually greatly accelerates tumorigenesis induced by expression of an oncogene (such as Myc; this applies to many other oncogenes, but not all). Tumors that arise inevitably have lost expression or function of the remaining wild-type p53 allele.

p53 is activated in cells by two different mechanisms. The first responds to oncogenes; the second responds to DNA damage (see above). As tumors lose p53, they become genomically unstable, accumulating mutations and other DNA abnormalities, including chromosomal translocations as well as gain or loss of chromosomes. Presumably, this reflects a failure of the p53-defective cells to respond to DNA damage by engaging efficient repair mechanisms or undergoing apoptosis.

NUCLEAR p53 INDUCES EXPRESSION OF A VARIETY OF GENES WITH DIFFERENT FUNCTIONS, INCLUDING APOPTOSIS

When p53 accumulates, it moves to the nucleus where it regulates a variety of genes. Here, we focus on those that control apoptosis, but it is useful to survey the other consequences as well. These are illustrated in Figure 11.9.

When p53 is activated, choices are made among the different consequences. The cell may die or it may arrest in the cell cycle and repair damage to the DNA. Different

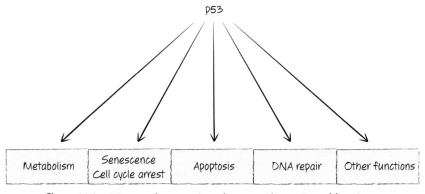

Figure 11.9. p53 controls expression of genes with a variety of functions.

tissues respond differently, and exactly what influences the decision is not well understood. It may well be that it is simply the level of p53 accumulation that determines the outcome; less p53 may engage repair, whereas more may lead to apoptosis.[8]

Several transcriptional targets of p53 promote cell death. Which of the pathways causes apoptosis triggered by p53 may depend on additional factors, including the tissue that is involved. One important apoptosis target of p53 is the BH3-only protein PUMA (its name means "p53–up-regulated mediator of apoptosis"), which binds to and inactivates all of the anti-apoptotic BCL-2 proteins (e.g., BCL-2, BCL-xL, MCL-1, and A1; see Chapter 5). It does not appear to activate BAX or BAK, although this is controversial, and additional signals may be required for MOMP and the mitochondrial pathway of apoptosis when p53 is stabilized (see below).

Mice lacking PUMA are resistant to DNA damage-induced p53-dependent apoptosis in the thymus (which is normally very sensitive to this). They show no increase in spontaneous cancer, although if oncogenes are expressed, cancer is somewhat accelerated, but not to the extent seen in mice with defective p53. Other p53 targets therefore presumably contribute to tumor suppression.

Another p53 target is the BH3-only protein NOXA. NOXA neutralizes the anti-apoptotic BCL-2 proteins MCL-1 and A1, but not BCL-2 or BCL-xL. However, lack of NOXA does not generally have dramatic effects on p53-mediated apoptosis. Nevertheless, NOXA may influence apoptosis (and tumor suppression) in some settings.

Yet another target is the BH3-only protein BID. This protein, when activated by proteolytic cleavage, can activate BAX and BAK to cause MOMP and engage the mitochondrial pathway (see Chapter 5). As we have discussed, BID is efficiently activated by caspase-8 following ligation of death receptors. For this reason, activation of p53 sensitizes cells to death receptor–induced apoptosis (see Chapter 6).

The pro-apoptotic BCL-2 effector BAX (see Chapter 5) is a direct target of p53 transcriptional activity in humans. In rodents, this is not the case, but stabilizing p53 nevertheless causes increased expression of BAX by an indirect mechanism. Although increased BAX expression does not itself cause MOMP, it does make cells more sensitive for engagement of this pathway. The activation of p53 therefore has a number of effects that promote the mitochondrial pathway of apoptosis (Fig. 11.10).

p53 also induces expression of the death receptors CD95 and DR5 (one of the TRAIL receptors). Together with the up-regulation of BID, this makes cells more sensitive to death ligands when p53 is activated. However, the extent to which the death receptor pathway contributes to p53-induced apoptosis is unclear. Mice lacking CD95 or its ligand show no defects in p53-mediated cell death, although mice that have de-

---

[8] This is a bit simplistic; p53 is modified in many ways that affect its transcriptional activity (through alterations in the binding of additional transcriptional cofactors). Many studies have suggested that the different outcomes of p53 activation depend on specific modifications of this protein.

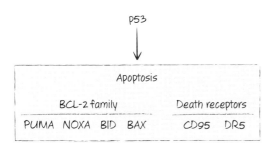

Figure 11.10. p53 controls a number of pro-apoptotic genes.

ficient TRAIL signaling show increased metastasis in a tumor model. It is therefore possible that loss of DR5 expression can contribute to the effects of loss of p53.

Which of the apoptotic pathways are most important for p53-induced cell death and how crucial is apoptosis for tumor suppression by p53? A hint comes from studies in which tumors are induced by the constitutive expression of Myc in B lymphocytes in mice. Lymphomas that arise in such animals appear slowly and generally have inactivating mutations, deletions, or loss of expression of p53. However, if lymphomas are induced by coexpressing Myc and BCL-2, which results in much more rapid tumorigenesis, the transformed cells do not develop mutations in p53. At least in these cells, inhibition of the mitochondrial pathway of apoptosis therefore appears to be sufficient to bypass the tumor-suppressive effects of p53. This further suggests that the mitochondrial pathway of apoptosis is a major mechanism of tumor supression (again, at least in these cells).

CYTOSOLIC p53 MAY ALSO FUNCTION IN APOPTOSIS

When p53 is stabilized and accumulates in cells, it can be found in both the nucleus and the cytosol. Cytosolic p53 can promote apoptosis by a mechanism other than regulation of transcription. Like some BH3-only proteins, it can directly activate the pro-apoptotic effector protein BAX (and, at higher concentrations, BAK) to cause MOMP. Moreover, as with BH3-only proteins, p53 is bound by anti-apoptotic BCL-2 proteins. Note that despite these activities, p53 does not have a BH3 domain and shares no sequence similarity with BCL-2 proteins.

As cytosolic p53 accumulates, it may be sequestered by anti-apoptotic BCL-2 proteins. When nuclear p53 accumulates, however, we have already seen that it induces expression of PUMA and BAX. PUMA is remarkably good at displacing p53 from anti-apoptotic BCL-2 proteins, releasing p53 to activate BAX and promote MOMP (Fig. 11.11).

Although biochemical and cellular studies support this cytosolic role for p53 in apoptosis, no compelling genetic evidence unambiguously shows that this functions in tumor suppression. It is a complicated problem, because without transcription factor

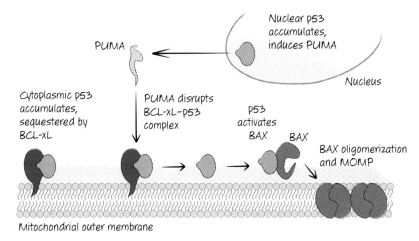

Figure 11.11. Nuclear and cytoplasmic p53 may cooperate in causing MOMP.

activity, p53 cannot induce expression of its regulator MDM2 or PUMA to promote the cytosolic activity. Clever experiments will be needed to resolve this controversial area of p53 research.

p53-INDUCED APOPTOSIS IN OTHER ANIMALS

Proteins related to the mammalian p53 family have been found not only in vertebrates but also in several invertebrate phyla. In *Drosophila*, a p53-like protein is activated in response to DNA damage. As in humans, *Drosophila* p53 acts as a transcription factor, and it leads to apoptosis. In the fly, it induces the expression of Reaper, Hid, and Sickle, all of which are direct transcriptional targets (see Fig. 11.12).

There are, however, fundamental differences between p53 in mammals and p53 in the fly. Although activation of p53 in mammalian cells leads to either cell cycle arrest or apoptosis (among other things), activation of p53 in the fly does not arrest cell division. In addition, the protein is regulated in a different way—there is no

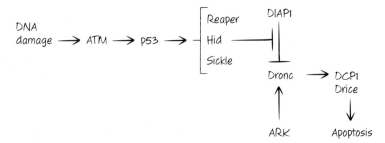

Figure 11.12. *Drosophila* p53 induces apoptosis.

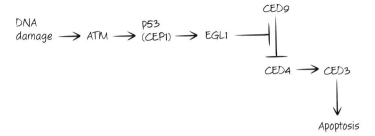

Figure 11.13. Nematode p53 induces apoptosis.

MDM2 homolog in insects. Intriguingly, *Drosophila* have a homolog of ATM, and this seems to regulate p53 in the fly.

Nematodes also have a p53 protein, which is called CEP1. Activation of CEP1 induces apoptosis, but like in flies, this does not cause cell cycle arrest. CEP1 is activated following DNA damage and transcriptionally up-regulates the BH3-only protein EGL1, leading to apoptosis (Fig. 11.13). Like in the fly, *Caenorhabditis elegans* does not have an MDM2 homolog but does have a homolog of ATM that regulates p53.

Clearly, p53 in these organisms is linked to the core apoptotic pathways that function in them. However, it is not obvious that this has tumor-suppressor functions, because the short life spans of such animals make cancer a minor selection force. It is likely that the role of p53 in these organisms is to preserve the genome in the germ line, so that cells that have extensive DNA damage die, rather than contribute to the next generation. The role of p53 in longer-lived invertebrates may be a different story, and p53-family members have been identified in other phyla. At present, we do not know what functions, if any, these have in tumor suppression.[9]

SOME COMPONENTS OF APOPTOTIC PATHWAYS ARE TUMOR SUPPRESSORS

Several proteins that we have already met are also tumor suppressors. Genetic ablation of BAX, BIM, or PUMA (see Chapter 5) promotes tumorigenesis in mice when a known oncogene, such as Myc, is expressed. In addition, PUMA and another pro-apoptotic BCL-2 protein, BOK, are frequently deleted in human cancers. Conversely, human tumors often amplify the genes encoding the anti-apoptotic proteins MCL-1 and BCL-xL.

APAF1, caspase-9, and caspase-3 have not been observed to act as tumor suppressors in mouse models. This is intriguing, because human cancers often have mu-

---

[9] In fact, p53 is part of a family of related transcription factors that includes p63 and p73 (collectively, the p53 family). p53 members present in flies and nematodes more closely resemble p63, which in vertebrates has roles in stem cell function (more than it has roles in apoptosis).

tations in caspase-3, indicating that it might be a tumor suppressor in humans. As we mentioned in Chapter 8, caspase-2 acts as a tumor suppressor in one model system, although we do not know whether this involves apoptosis. Another interesting example is caspase-8, which is often mutated in some types of human cancer, such as neuroblastoma.

OTHER TUMOR SUPPRESSORS ENGAGE APOPTOTIC PATHWAYS

Many tumor suppressors influence apoptosis indirectly. After p53, the most common tumor suppressor that is mutated in cancers is PTEN. PTEN is an enzyme that removes a phosphate group from phosphatidylinositol-3,4,5-*tris*-phosphate (PIP_3), a lipid involved in activation of AKT. AKT promotes cell growth and inhibits apoptosis by several mechanisms (see Chapter 5). By blocking AKT activation, PTEN can inhibit cell proliferation and promote apoptosis. When PTEN is inactivated by mutation, AKT is constitutively active, and this promotes cancer.

There are many other well-defined tumor-suppressor pathways, and in several cases, it is clear that they influence apoptosis. Their links to specific apoptotic pathways are less well established, however, and we have much to learn about how they control apoptosis.

APOPTOSIS AS AN ACHILLES HEEL OF CANCER

At first glance, any disruption of apoptosis that occurs as a step in oncogenesis might seem to make killing a cancer an intractable problem. But this is not the case. Consider what it takes to make a tissue expand rather than regress. If the rate of cell division is greater than the rate of cell death, cells accumulate (the tissue/cancer expands); conversely, if the rate of cell death is greater than the rate of cell division, the tissue (or cancer) regresses. This means that *anything* that tips the balance toward cell division over cell death will result in the expansion of a tissue or cancer. Tumor suppressors, whether they act by inhibiting the cell cycle or promoting cell death, push the balance the other way, and in general, this works to prevent cancer.

But when cancer occurs, mutations only need to be sufficient to tip the scale toward expansion. Apoptosis mechanisms may be engaged—for example, in response to tumor suppressors that are activated or apoptotic signals from oncogenes such as Myc—but these do not effectively tip the balance in our simple equation. Thus, in many cases, cancer cells are essentially "primed" for death but do not die in sufficient numbers to prevent the expansion of a tumor. The goal is therefore to devise therapies that take advantage of such priming and tip the balance toward regression.

Remember one way in which apoptotic pathways are controlled: Activation of BAX and/or BAK induces MOMP unless anti-apoptotic BCL-2 proteins prevent this by

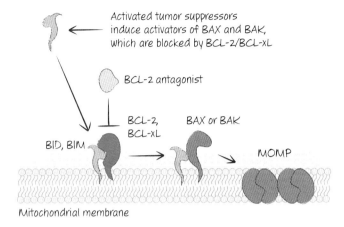

Figure 11.14. One way a BCL-2 antagonist can kill a cell.

sequestering BH3-only proteins (such as BIM) or activated BAX and BAK themselves. If a drug could disrupt the interaction between pro-apoptotic BCL-2 proteins and anti-apoptotic BCL-2 proteins, apoptosis should proceed (Fig. 11.14). Indeed, a number of such drugs are currently in clinical trials.

But what would happen if a normal cell were treated with a BCL-2 antagonist? Without tumor-suppressor signals to activate BAX and BAK (such signaling would not be occurring in normal cells), neutralizing the anti-apoptotic BCL-2 protein would not be sufficient to engage MOMP. In contrast, the primed cancer-causing cells would be more prone to engaging the mitochondrial pathway, and only they would die. One could therefore preferentially destroy the cancer cells.

OTHER FORMS OF CELL DEATH IN CANCER

Cancers are not well-organized tissues, and as they grow, they expand beyond their ability to provide nutrients and oxygen to all of the cells in the tumor. As a result, cancers often have central regions of extensive cell death that are necrotic (see Chapter 8). Although such cell death certainly fits into our simple equation above, necrosis in a cancer is often associated with a poor prognosis. This may be because it activates inflammatory responses (see Chapter 9) that produce cytokines that stimulate growth of the cancer. Nevertheless, tipping the balance in favor of any form of cell death will produce regression, and there are therapies that aim for such an outcome.

Autophagy is usually a cell survival mechanism, and inhibition of autophagy is another goal in tumor therapy. But Beclin-1, which initiates autophagy (see Chapter 8), is a tumor suppressor that is frequently deleted in human cancer. Mice that lack one allele of Beclin-1 display spontaneous cancer and show accelerated tumor formation when an oncogene is expressed. Cells that have defective autophagy accumulate unfolded proteins, and these induce the production of reactive oxygen species that can promote additional mutations that contribute to oncogenesis.

Induction of autophagic cell death (see Chapter 9) is also being explored as a therapy. Chloroquine (which has a long history in the treatment of malaria infection) causes autophagic cell death, and several trials with this drug in combination with other therapeutics are under way.

The regulation of cell death in cancer and cancer therapy is varied and complex. Tumor-suppressor mechanisms can specify cells for death, much as do developmental cues (Chapter 10), and oncogenesis necessarily involves mechanisms to avoid such cell death. An important step forward is to identify ways to counteract these avoidance mechanisms.

Of course, cancer is not the only disease process in which cell death pathways have a prominent role. The approaches to controlling cell death and survival will have applications in the treatment of autoimmunity, neurodegenerative diseases, and aging. In Chapter 12, we consider how our knowledge of cell death processes can be tested by formal and practical approaches.

Chapter 12

The Future of Death

TESTING CELL DEATH MODELS

So far, we have surveyed a great deal of information, organizing it into superficially complete signaling pathways that lead from a variety of stresses and developmental cues to the death of a cell. We understand, in principle, how a cell is specified for death, how this engages cell death pathways, how these are regulated, how death is carried out, how the cell is removed, and the consequences for the cells that remain. It is tempting to think that what remains to be discovered is merely the details, mainly of interest to specialists. But is this "big picture" really complete, and are there approaches we can use to determine this?

In this final chapter, we consider two such approaches to this problem, one formal and one informal. The first involves taking what we know, modeling it mathematically, and asking whether the models we generate conform to what we observe. This approach does not explicitly tell us if we are right or wrong but can reveal areas of ignorance and provide a different way to look at the pathways that we have considered.

The second approach is a practical one: Can we use the information we have gleaned to successfully engineer cell death? Again, this cannot tell us whether we are right or wrong but, instead, where our ignorance lies and if our knowledge is useful.

CELLULAR STATES AND MODELS OF DEATH

Living cells can be in many states, including rest, proliferation, differentiation, and various functions and can transition between such states. But any transition to the cell death state is irreversible. The consequences may differ, but from the point of view of the cell, dead is dead.

Apoptosis (as well as most other forms of cell death) is a balancing act between molecular mechanisms that determines whether a cell survives or dies. The outcome depends on the relative amounts of specific molecules in the cell and the interactions between them. A simple experiment shows that cell-to-cell variation in these molecules is stochastic. If cells are cultured with pro-apoptotic signals and monitored (e.g., by time-lapse microscopy), we see that some cells die relatively quickly, some slowly, and, if the signal was not too overwhelming, some survive. Which cells do what appears to be random. However, if a cell divides and one of the daughters dies, the other daughter is more likely to die in a given time frame than are other cells that are less closely related (Fig. 12.1). This correlation is strongest immediately after division and wanes with time. Undoubtedly, this is because the quantities of key molecules in the daughter cells are more similar than the quantities of such molecules in more distantly related cells, but as it happens, no one molecule accounts for the variation. Even when all of the cells used are derived from a single cell, the divergence over generations is remarkably rapid. This means that there is significant "noise" in the system.

The simple idea that levels of key molecules randomly vary between cells and that the outcome of a signaling event depends on these levels applies to many cellular responses, not just cell death. But cell death is a special case, as we have said, because it is irreversible. Given a constant level of stress (and our cells are always under some stress) and random variation in the levels of different key molecules, all cells would eventually die (for some cells, this does not actually occur—neurons and cardiac myocytes, for example, are postmitotic and most persist throughout the life span of the individual). In other cases, random cell death can be offset by proliferation, both in normal tissue (see Chapter 10) and neoplastic tissue (see Chapter 11). The trick for

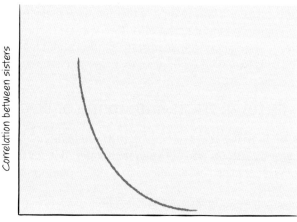

Figure 12.1. Cell division and stochastic death. If following cell division, sisters cells are monitored for their response to an apoptotic stimulus, MOMP is closely correlated shortly after division, but this relationship quickly decays with time.

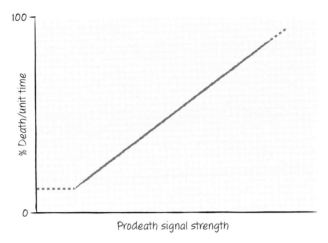

Figure 12.2. Expected relationship between the strength of a prodeath signal and outcome, if regulatory mechanisms are absent.

complex biological systems is to control this balance and limit the variability through regulation. One way to look at this is to consider the probability that a given cell will die in response to a given level of stress (or any prodeath signal) during a fixed period. In the absence of built-in regulatory mechanisms, the relationship would be expected to be largely linear (Fig. 12.2).

But the actual, observed relationship between stress and cell death is more similar to that shown in Figure 12.3. This type of relationship, which approximates a step function, is all around us. It is in biological systems of all types, but also in many other

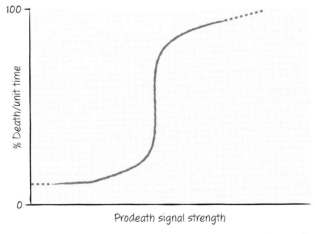

Figure 12.3. Observable relationship between the strength of a prodeath signal and outcome.

systems. Often, it is an effect of negative feedback regulation.[1] But cell death pathways (those that we understand) do not have any obvious built-in negative feedback mechanisms. If we look back on the pathways that we have covered, despite the existence of many inhibitory interactions, there appear to be no points at which the induction of a later event inhibits an earlier event in the pathway. How then is cell death controlled to produce this step function?

BISTABILITY IN CELL DEATH PATHWAYS

A system that can exist in two stable states (e.g., survival vs. death, inactive vs. active caspases, or intact mitochondria vs. permeabilized mitochondria) is bistable.[2] We can analyze such systems by drawing a pathway, assigning concentrations (real or estimated) for all of the components, and then generating mass-action formulae. A stimulus is introduced (simulated by the increase of an upstream component over a range) and the resulting curves can be generated (usually in silico). We can then see if our model behaves in a bistable manner, displaying two states. A robust bistable system is one that is relatively resistant to small changes in concentrations of the different components (noise).

Let us begin with an example discussed in Chapter 5. Recall that there are two models for how interactions among the BCL-2 proteins lead to MOMP in vertebrate cells (Fig. 12.4): (1) the neutralization model, where inhibition of anti-apoptotic function by BH3-only proteins is necessary and sufficient to cause MOMP, which is mediated by the pro-apoptotic BCL-2 effector proteins, and (2) the more complex direct activator/derepressor model, in which direct activators for the effectors are se-

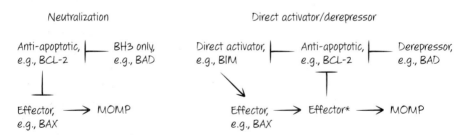

Figure 12.4. Schematic representation of two models of BCL-2–family protein interactions in MOMP. (Asterisk) Activated effector.

---

[1] This is the insight used by Robert Fulton in creating the steam engine.

[2] We use the term bistable to mean "having two states that can stably exist as input changes." Technically, this is incorrect: Formally, "bistability" is a consequence of a mathematical representation having two solutions. In its use here, it is more correctly called *transcritical bifurcation*, which can be bistable or not, in the formal sense. Our admittedly sloppy use of *bistability* in this context is taken from literature in this area. As for our use of the term "stable," we refer to a state that can persist despite noise in the system; e.g., death is a persistent state (of course), but so too must cell survival be if we are to function as organisms.

questered by the anti-apoptotic BCL-2 proteins, and then released to trigger the effectors, leading to MOMP.

When tested as described above, the simpler neutralization model is unstable. The direct activator/derepressor model is bistable but has a problem: It is not very robust. That is, when the system is subjected to noise (more than a very small amount of variation in the component molecules), it is unstable.

Does this mean that neither model is correct? Not necessarily. The starting concentrations used for the components may have been wrong in our model, or the model may simply need tweaking. It turns out that a small tweak to the second model makes it work in a robustly bistable manner (showing much more resistance to noise): The tweak is the positive feedback loop shown in Figure 12.5.

Now, the activated pro-apoptotic BCL-2 effectors (BAX and BAK) can activate inactive effectors, without relying on the appearance of more direct activator proteins. Indeed, this is supported by experimental evidence. Intuitively, one might have thought that this addition to the system would have made it even less resistant to random fluctuations, but the results show otherwise.

This does not show that the direct activator/derepressor model is correct. But the pathway is formally consistent with robust bistability: It is an on-off switch that is resistant to random noise, such as small variations in the amounts of the main components. The model itself still may not be correct or only partially so. For example, it does not include inhibition of the activated pro-apoptotic effectors BAX and BAK by the anti-apoptotic BCL-2 proteins, and we have good reason to believe that this occurs. Would adding such interactions make the system more or less robust? At this point, we do not know.

MOMP is not the only source of bistability in apoptosis, and the mitochondrial pathway is only one route to cell death. Bistability is also found in caspase activation and inhibition by IAPs (see Chapter 3). It emerges if we assume that although IAPs can inhibit caspases, caspases also inhibit IAPs (that is, the bound caspase prevents the IAP from inhibiting any other caspases, which is a reasonable assumption).

Another source of bistability in cell death occurs at the level of transcription. As discussed in Chapter 11, p53 induces the expression of a number of proteins with various functions, including apoptosis. Among the proteins induced by p53 is its inhibitor MDM2, and the p53-MDM2 interaction and its regulation have been shown to be bistable.

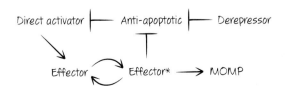

Figure 12.5. Schematic representation of the direct activator/derepressor model of BCL-2 protein interactions in MOMP. Included is a feed-forward mechanism, whereby activated effector proteins can activate other effector proteins.

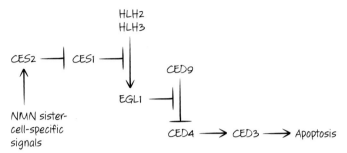

Figure 12.6. NMN neuron pathway of apoptosis in nematodes.

IDENTIFYING OTHER LIFE–DEATH DECISIONS

Some of the cell death pathways that we have discussed are not inherently bistable, even though the result (e.g., life vs. death) clearly achieves this. An example is the apoptotic pathway in the NMN neurons in *Caenorhabditis elegans* development, discussed in Chapter 10 and shown again in Figure 12.6.

Although this has not been formally tested for bistability, the pathway, as we understand it, is linear and probably insufficient for robust bistability. If so, then from where does bistability in the system arise? The same query applies to the other nematode cell deaths that occur during development and in response to genotoxic stress in germ cells (see Chapters 10 and 11). It is likely that bistability arises from upstream transcriptional control mechanisms, whereas the pathway itself (composed of several "inhibitor of inhibitor" connections) functions to dampen the effects of noise in the system. If this proves to be correct (following formal analysis of the pathway), with this information we might set out to identify the transcriptional control mechanisms that convert the pathways to bistable systems.

This fresh perspective can be brought to other apoptotic pathways as well. Developmental cell death in *Drosophila* (see Chapter 10) includes the pathway shown in Figure 12.7. Again, this pathway has not been formally tested for bistability, although one source may reside within the IAP–caspase interactions, as discussed above. However, as with developmental cell death in nematodes, it depends on transcriptional control of the IAP antagonists Reaper, Hid, and Grim. We saw in Chapter 10 that during metamorphosis, such transcriptional control can be triggered by the binding of

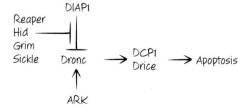

Figure 12.7. Developmental apoptosis in *Drosophila*.

ecdysone to its nuclear receptors. Bistability may emerge from tissue-specific regulatory interactions governing the expression and function of the ecdysone receptor and/or other regulatory steps in the pathway that remain to be uncovered.

CELL DEATH BY DESIGN

The other approach to evaluating our understanding of cell death—one perhaps more in keeping with the mandate placed on us by the public and private entities that support the research enterprise—is to ask if we can manipulate the system for desired ends. There is obviously great therapeutic potential to controlling the survival and death of specific cells at will. But to do this, we must probe each step in the cell death mechanisms that we have detailed to determine where the effective points of manipulation lie. When we consider such manipulations, we are especially concerned with two approaches: pharmacology and genetic engineering. Both hold promise. But as we will see, there is a surprising amount that we do not yet know.

MAKING CELLS DIE

It is easy to kill a cell; the trick is to kill only those cells we wish to, with as little collateral damage as possible to other cells. A good place to start is the generation of "suicide switches."

Gene therapy, in principle, holds great potential for treatment of disease, but any introduction of foreign DNA into an individual's cells carries with it the risk of mutations that could lead to cancer.[3] One strategy is to include a switch that can trigger death in engineered cells at will. Some such suicide switches exploit the cell death pathways already present in our cells.

Several proteins have been identified that bind to the immunosuppressive drug FK506, collectively called FK-binding proteins (FKBPs). Sequences derived from one of these, FKBP12, can be fused onto a protein of choice, and a derivative of the drug (without immunosuppressive activity) can be safely used to dimerize chimeric molecules in cells (Fig. 12.8).

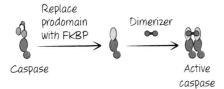

Figure 12.8. FKBP dimerization.

---

[3] Unfortunately, this is only one of many challenges facing gene therapy; the rest lie outside the scope of our discussion.

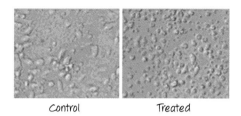

Control   Treated

Figure 12.9. A caspase-9 suicide switch in engineered T cells. Addition of the dimerizer agent (treated) induces rapid apoptosis in most of the cells.

Creating a suicide switch this way is straightforward (at least in principle). Because initiator caspases are activated by induced proximity by adapter molecules (see Chapter 3), we can replace the prodomain of an initiator caspase with the FKBP motif and express the chimeric protein in the cells that we wish to control. Addition of the FK506 derivative would then kill the cell. Figure 12.9 shows this for a caspase-9-based switch.

Whether such initiator caspase-based suicide switches are sufficiently resistant to the effects of noise to be of practical value is unclear. That is, we do not know if they behave as robust bistable systems. If not, additional elements will have to be incorporated until they do. Alternatively, this may not be the ideal point in the pathway to exercise control.

There are other potential therapeutic applications for drugs that promote cell death. Probably the most pressing one relates to cancer therapy. In Chapter 11, we discussed how cell death pathways can be thwarted in cancer, but we developed the idea that this can produce an "Achilles heel" when these pathways are engaged but held in check (i.e., cancer cells are often poised to die in a way that normal cells are not). The challenge is to disrupt the resistance mechanism and thereby kill the transformed cells.

Anti-apoptotic BCL-2 proteins block apoptosis by sequestering activated pro-apoptotic effectors (BAX and BAK) and the proteins that activated them. BH3-only proteins can disrupt this sequestration, leading to death (see Chapter 5 and above). We can mimic this effect using drugs that bind to the anti-apoptotic proteins in the same manner. One such drug is ABT-737 (or its orally available form, ABT-263), which binds to BCL-2, BCL-xL, and Bcl-w but not MCL-1 (Fig. 12.10).

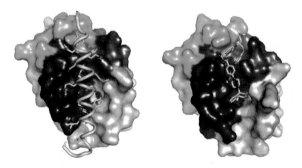

Figure 12.10. ABT-737 binds to the BH groove of BCL-xL in the same manner as do BH3-only proteins.

THE FUTURE OF DEATH 185

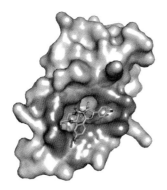

Figure 12.11. Nutlin-3 (green) bound to MDM2, (brown) binding pocket in MDM2 that binds to p53.

ABT-737 does not trigger death in most cells, with the exception of some lymphocytes and platelets (remarkably, the latter return to normal levels even during the course of treatment, perhaps by expressing MCL-1). In contrast, some tumor cells are very sensitive to the drug, rapidly undergoing apoptosis. Others are highly responsive to ABT-737 in combination with other chemotherapeutic agents.

Other survival mechanisms can also be targeted pharmacologically. In Chapter 11, we discussed how p53 is stabilized by disruption of the p53-MDM2 interaction, leading to apoptosis (and additional effects of p53). Although many tumors mutate p53 itself to bypass this control mechanism, many others subvert the p53 pathway by losing ARF (which normally blocks the p53-MDM2 interaction). In the latter case, p53 can be stabilized by drugs that disrupt the binding of p53 to MDM2. One such drug is nutlin-3 (Fig. 12.11). At this point, we do not know if this is effective as an anticancer therapy but the prospect is intriguing.

These are only two examples of pharmacological approaches to making cells die. Others involve blocking survival signals or engaging death pathways upstream of the core molecular mechanisms. Indeed, most cancer therapies are designed to kill tumor cells, and to do so they must engage one or more of the cell death pathways that we have discussed.

MAKING CELLS LIVE

Just as it can be desirable to make some cells die, it can be beneficial to limit cell death in some cases. The approach taken depends on the mode of cell death and the pathways with which it is engaged.

Caspases (like many proteases) are amenable to pharmacological inhibition (see Chapter 2). However, inhibition of caspases does not necessarily prevent cell death. Remember, two pathways of caspase activation and apoptosis result in cell death even when caspases are blocked: (1) death-receptor-mediated engagement of RIPK1-dependent necrosis and (2) MOMP-dependent caspase-independent cell death. Nevertheless, in some cases, inhibition of caspases does, indeed, preserve the survival of

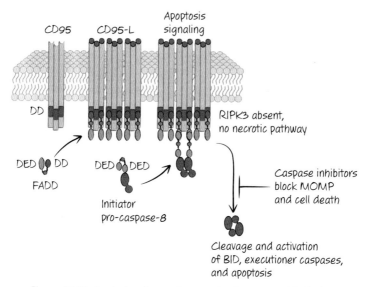

Figure 12.12. Fas-L signaling and protection by caspase blockade.

the cell. As discussed in Chapter 6, caspase inhibitors preserve hepatocytes and the survival of animals injected with CD95-ligand or antibodies that stimulate the death receptor CD95 (Fig. 12.12). This is because the RIPK1-dependent necrosis pathway is not active in these cells.

This has potential value in the treatment of conditions that cause liver pathology via CD95-ligand-mediated apoptosis. These include chronic hepatitis infection and alcoholic cirrhosis. As it turns out, all caspase inhibitors tested in humans so far tend to accumulate in the liver, which may be useful for these conditions (but more problematic for other target tissues). That said, the long-term consequences of caspase inhibition remain unknown.

Other examples of cell death that depend on caspase activity include cell death caused by caspase-1 activity and, potentially, death that is dependent on the activation of caspase-2 (see Chapter 7). Lethal bacterial sepsis in experimental animals that involves the former has been effectively treated with caspase inhibitors. Whether such approaches will have therapeutic value will depend on the development of new caspase inhibitors with better availability in vivo.

RIPK-dependent programmed necrosis can be engaged by death receptors and probably other mechanisms (see Chapter 8). RIPK1 has been shown, in principle, to be an effective target for pharmacological inhibition by necrostatins, and whether RIPK3 (also required for this death pathway) can be specifically targeted remains to be seen. At this point, necrostatins have not been developed for human use, but the potential is there: Mice lacking RIPK3 show resistance to necrotic injury of the pancreatic islet cells in a model of acute pancreatitis (Fig. 12.13).

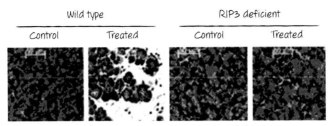

Figure 12.13. RIPK3 and pancreatic injury. Wild-type or RIPK3-deficient mice were treated with an agent that induces necrotic injury in the pancreas. Animals without RIPK3 were protected.

In Chapter 8, we discussed the role of the mitochondrial permeability transition (MPT) in necrosis, for example, in ischemia/reperfusion injury. Mice lacking cyclophilin D display a defective MPT and are resistant to such injury. The drug cyclosporin A blocks the MPT by inhibiting cyclophilin D. Although this drug has other effects (it is widely used as an immunosuppressive drug in organ transplant, at doses well below those needed to inhibit MPT), other compounds that more specifically inhibit cyclophilin D have been described. These drugs may have promise in the treatment of stroke and myocardial infarction as well as other diseases involving ischemia/reperfusion injury.[4]

As another approach to keeping cells alive, we would ideally like to have drugs that inhibit the activation of BAX and BAK to prevent MOMP and all forms of MOMP-dependent cell death. At present, we do not sufficiently understand BAX/BAK activation and function to design such agents from first principles. However, we can stimulate pathways that induce the expression or stabilization of anti-apoptotic proteins and thereby prevent MOMP to preserve cell survival. One example is a pharmacological inhibitor of GSK3 that promotes cell survival, at least in part, by extending the half-life of MCL-1 (Fig. 12.14). We can envision using such drugs to limit damage to tissues such as the gut during chemotherapy for cancer, provided the agents can be delivered to normal tissue without compromising their antitumor effects.

BACK TO THE FUTURE

There are thus ways to challenge our knowledge of cell death by probing models of the pathways in silico and by engineering the systems for applications in vivo. Beyond these, it may be tempting simply to stand back and admire the pool of understanding that we have gained. But when we do, it is only necessary to extend our gaze further to view a sea of ignorance of which our knowledge only skims the surface.

An elegant way to illustrate this is by considering the "button experiment."[5] Imag-

---

[4] Although cyclosporin A has protective effects in animal models of ischemia/reperfusion injury, neurotoxic effects have been observed in up to 60% of organ transplant patients who receive this drug as an immunosuppressant. The high doses needed to block cyclophilin D are therefore problematic for this drug.

[5] Conceived by Stephan Kaufmann, a theoretical biologist and geneticist.

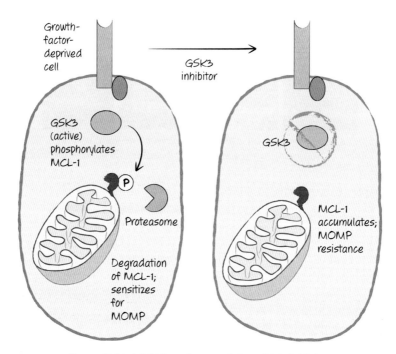

Figure 12.14. Inhibition of apoptosis by a GSK3 inhibitor.

ine a table covered with buttons. Choose, at random, two buttons and connect them with a piece of thread. Then choose another button, again at random, pick it up, and count how many buttons are collected with it. Now repeat the entire process and score again. If we do this many times (this is more practically done in a computer model) something interesting happens (Fig. 12.15).

Early on, one or at most a few buttons come up with the one we pick up. Later, the entire system undergoes a phase transition such that any chosen button brings with it a large number of other buttons. Quickly, nearly any chosen button pulls with it most of the others. Astonishingly, this transition occurs near the point at which the number of threads approximately equals half the number of buttons. Chosen randomly, some buttons act as "nodes," tied to many threads, and some remain attached to only one or two. But at this phase transition, the system has been altered dramatically.

The human genome is composed of about 28,000 genes encoding proteins. We take it as a given that all of these *do something*. The majority are acted on by and/or act on at least one other protein. But proteins are not the only "buttons." Lipids, carbohydrates, nucleic acids (and their sequences), and metabolites are all buttons, similarly connected by threads. Viruses, bacteria, and other passengers in our bodies carry

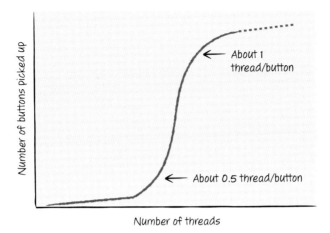

Figure 12.15. The "button experiment."

their own sets of buttons and threads, and many of these connect to our buttons as well. It is a dizzying thought that picking up almost *any* button will drag along large numbers of other buttons.

Clearly, any button that we choose is likely to be connected to the pathways controlling life and death. Perturb a protein, alter a metabolite, mutate a sequence, add a drug, and we can impact cell death pathways. Because the cell death pathways are robustly bistable, most perturbations will not shift the state dramatically, but they have the potential to do so. Selection has acted over eons to favor those buttons and threads that impart fitness to the pathways for cell death and survival. Moreover, the molecular events that control energy, movement, replication, repair, and all other important cellular processes connect to this machinery. The devil is in the details: *How are they all connected?* This is the future to which we will return.

Darwin concluded *On the Origin of Species* with an evocative image: an "entangled bank" on which varieties of life forms comingled, competed, and survived. We now know that evolution acts not only at the level of individuals, but also at the level of cells engaged in development, homeostasis, and disease. Cell survival and death are the outcomes of battles on the entangled banks within all of us.

Figure Credits

Chapter 1: **1.1**, Images provided by Douglas R. Green. **1.2**, Reprinted from Maclean KH, et al. 2008. *J Clin Invest* **118:** 79–88, ©2008 with permission from the American Society for Clinical Investigation. **1.3**, Images courtesy of Dr. Nigel Waterhouse, Mater Medical Research Institute, Brisbane, Australia.

Chapter 2: **2.9–2.11**, Reprinted from Mahrus S, et al. 2008. *Cell* **134:** 866–876, ©2008 with permission from Elsevier. **2.12**, Image courtesy of Dr. Yufang Shi, Dept. of Immunology, University of Medicine and Dentistry of New Jersey; structure (*inset*), PDB 1GQF (Riedl S, Bode W, Fuentes-Prior P. 2001. *Proc Natl Acad Sci* **98:** 14790). **2.13**, Reprinted from Ellis HM, Horvitz HR. 1986. *Cell* **44:** 817–829, ©1986 with permission from Elsevier. **2.14**, Reprinted from Quinn LM, et al. 2000. *J Biol Chem* **275:** 40416–40424, ©2000 with permission from the American Society for Biochemistry and Molecular Biology. **2.15**, Reprinted with permission from Macmillan Publishers Ltd.: Kuida K, et al. 1996. *Nature* **384:** 368–372, ©1996.

Chapter 3: **3.1a**, PDB 1K86 (Chai J, Wu Q, Shiozaki E, Srinivasula SM, Alnemri ES, Shi Y. 2001. *Cell* **107:** 299–407); **b**, PDB 1GQF (Riedl S, Bode W, Fuentes-Prior P. 2001. *Proc Natl Acad Sci* **98:** 14790). **3.2, left**, *see* 3.1a; **center**, PDB 1F1J (Wei Y, Fox T, Chambers SP, Sintchak J, Coll JT, Golec JM, Swenson L, Wilson KP, Charifson PS. 2000. *Chem Biol* **7:** 423–432); **right**, *see* 3.1b. **3.3**, Reproduced from Paul WE. 2008. *Fundamental immunology*, 6th Edition, ©2008 with permission from Lippincott, Williams & Wilkins. Image kindly provided by Drs. Ing Swie Goping and Chris Bleackley, University of Alberta, Edmonton. **3.7, left to right**, PDB 1DDF (Huang B, Eberstadt M, Olejniczak ET, Meadows RP, Fesik SW 1996. *Nature* **384:** 638–641); PDB 1CWW (Day CL, Dupont C, Lackmann M, Vaux DL, Hinds MG. 1999. *Cell Death Differ* **6:** 1125–1132); PDB 1PN5 (Hiller S, Kohl A, Fiorito F, Herrmann T, Wider G, Tschopp J, Grutter MG, Wuthrich K. 2003. *Structure* **11:** 1199–1205); PDB 1A1Z (Eberstadt M, Huang B, Chen Z, Meadows RP, Ng SC, Zheng L, Lenardo MJ, Fesik SW. 1998. *Nature* **392:** 941–945). **3.8**, Reproduced with permission from Fuentes-Prior P, Salvesen GS. 2004. *Biochem J* **384:** 201–232, ©The Biochemical Society. **3.9**, Reprinted by permission from Macmillan Publishers Ltd.: Fernandes-Alnemri T, et al. 2007. *Cell Death Differ* **14:** 1590–1604, ©2007. **3.12**, PDB 1CP3 (Mittl PR, Di Marco S, Krebs JF, Bai X, Karanewsky DS, Priestle JP, Tomaselli KJ, Grutter MG. 1997. *J Biol Chem* **272:** 6539–6547). **3.13**, Images courtesy of Dr. Nigel Waterhouse, Mater Medical Research Institute, Brisbane, Australia. **3.14**, Reprinted from Wang SL, et al. 1999. *Cell* **98:** 453–463, ©1999 with permission from Elsevier. **3.16**, PDB 1I3O (Riedl SJ, Renatus M, Schwarzenbacher R, Zhou Q, Sun C, Fesik SW, Liddington RC, Salvesen GS. 2001. *Cell* **104:** 791–800).

Chapter 4: **4.2**, PDB 3YGS (Qin H, Srinivasula SM, Wu G, Fernandes-Alnemri T, Alnemri ES, Shi Y. 1999. *Nature* **399:** 549–557). **4.5**, Images courtesy of Dr. Christopher Dillon, St. Jude Children's Research Hospital, Memphis, Tennessee. **4.6**, Images provided by Dr. Stephen Tait, St. Jude's Children's Research Hospital, Memphis, Tennessee. **4.8**, Reprinted from Chautan M, et al. 1999. *Curr Biol* **9:** 967–970, ©1999 with permission from Elsevier.

Chapter 5: **5.2**, Courtesy of Dr. Stephen Tait, St. Jude's Children's Research Hospital, Memphis, Tennessee. **5.3, left**, PDB 1MAZ (Muchmore SW, Sattler M, Liang H, Meadows RP, Harlan JE, Yoon HS, Nettesheim D, Chang BS,

---

Structures created using data from Research Collaboratory for Structural Bioinformatics Protein Data Bank (RCSB PDB; Berman HM et al. 2000. *Nucleic Acids Res* **28:** 235–242; http://www.pdb.org) and PyMOL (http://www.pymol.org/).
(AAAS) American Association for the Advancement of Science.

Thompson CB, Wong SL, et al. 1996. *Nature* **381:** 335–341); **right**, PDB 1BXL (Sattler M, Liang H, Nettesheim D, Meadows RP, Harlan JE, Eberstadt M, Yoon HS, Shuker SB, Chang BS, Minn AJ, et al. 1997. *Science* **275:** 983–986). **5.4, left**, PDB 1F16 (Suzuki M, Youle RJ, Tjandra N. 2000. *Cell* **103:** 645–654); **right**, PDB 2IMS (Moldoveanu T, Liu Q, Tocilj A, Watson M, Shore G, Gehring K. 2006. *Mol. Cell* **24:** 677–688). **5.11**, see 5.4, left; **overlay**, PDB 2K7W (Gavathiotis E, Suzuki M, Davis ML, Pitter K, Bird GH, Katz SG, Tu HC, Kim H, Cheng EH, Tjandra N, et al. 2008. *Nature* **455:** 1076–1081). **5.15**, PDB 2BID (Chou JJ, Li H, Salvesen GS, Yuan J, Wagner G. 1999. *Cell* **96:** 615–624). **5.24**, Reprinted from Hardwick JM, Youle RJ. 2009. *Cell* **138:** 404, ©2009 with permission from Elsevier. **5.25, left**, see 5.3, left; **right**, PDB 1K3K (Huang Q, Petros AM, Virgin HW, Fesik SW, Olejniczak ET. 2002. *Proc Natl Acad Sci* **99:** 3428–3433). **5.26, top left**, see 5.4, left; **top right**, PDB 1F0L (Choe S, Bennett M, Fujii G, Curmi PM, Kantardjieff KA, Collier RJ, Eisenberg D. 1992. *Nature* **357:** 216–222); **lower left**, PDB 1CII (Wiener M, Freymann D, Ghosh P, Stroud RM. 1997. *Nature* **385:** 461–464); **lower right**, PDB 2C9K (Boonserm P, Mo M, Angsuthanasombat C, Lescar J. 2006. *J Bacteriol* **188:** 3391). **5.27**, Image courtesy of Paul A. Ney, MD, St. Jude Children's Research Hospital, Memphis, Tennessee. **5.29**, PDB 2P1L (Oberstein A, Jeffrey PD, Shi Y. 2007. *J Biol Chem* **282:** 13123–13132).

Chapter 6: 6.2, PDB 1TNR (Banner DW, D'Arcy A, Janes W, Gentz R, Schoenfeld HJ, Broger C, Loetscher H, Lesslauer W. 1993. *Cell* **73:** 431–445). **6.4** and **6.5**, Redrawn from Wang L, Yang JK, Kabaleeswaran V, Rice AJ, Cruz AC, Park AY, Yin Q, Damko E, Jang SB, Raunser S, et al. The Fas-FADD death domain complex structure reveals the basis of DISC assembly and disease mutations. *Nat Struct Biol* (in press). PDB kindly provided by Dr. Hao Wu. **6.8**, Image courtesy of Dr. Ralph C. Budd, University of Vermont, Burlington.

Chapter 7: 7.6, left, Reprinted from Gillray J. 1799. *The Gout*. **7.13, left**, PDB 2OF5 (Park HH, Logette E, Raunser S, Cuenin L, Walz T, Tschopp J, Wu H. 2007. *Cell* **128:** 533–546); **right**, Reprinted from Park H, et al. 2007. *Cell* **128:** 533–546, ©2007 with permission from Elsevier.

Chapter 8: 8.1, Reprinted from Edinger AL, Thompson CB. 2004. *Curr Opin Cell Biol* **16:** 663–669, ©2004 with permission from Elsevier. **8.2**, Reprinted with permission from Springer Science + Business Media: Silva M, et al. 2008. *Apoptosis* **13:** 463–482, ©2008. **8.8**, Reprinted from Eskelinen E-L. 2008. *Intl Rev Cell Mol Biol* **266:** 207–247, ©2008 with permission from Elsevier. **8.13**, Reprinted from Lum J, et al. 2005. *Cell* **120:** 237–248, ©2005 with permission from Elsevier. **8.15**, Reprinted from Kirsch D, et al. 2010. *Science* **327:** 593–596, ©2010 with permission from AAAS.

Chapter 9: 9.2, Reprinted from Brumatti G, et al. 2008. *Methods* **44:** 235–240, ©2008 with permission from Elsevier. **9.4** and **9.11**, Reprinted from Hanayama R, et al. 2004. *Science* **304:** 1147–1150, ©2004 with permission from AAAS. **9.22**, Reprinted from Park J, et al. 2007. *Diagn Cytopathol* **35:** 806–809, ©2007 with permission from John Wiley and Sons.

Chapter 10: 10.1, Reprinted from Sulston JE, et al. 1983. *Dev Biol* **100:** 64–119, ©1983 with permission from Elsevier. **10.2**, Reprinted from Abraham M, et al. 1007. *Dev Cell* **12:** 73–86, ©2007 with permission from Elsevier. **10.5**, Reprinted from Maurer CW, et al. 2007. *Development* **134:** 1357–1368, ©2007 with permission from The Company of Biologists. **10.7**, Reprinted from Lee C-Y, et al. 2001. *Development* **128:** 1443–1455, ©2001 with permission from The Company of Biologists. **10.8**, Reprinted from Neufeld T, Baehrecke E. 2008. *Autophagy* **4:** 557–562, ©Landes Bioscience. **10.9, left**, Reproduced with permission from Zuzarte-Luís V, Hurlé JM. 2002. *Int J Dev Biol* **46:** 871–876; **right**, Reprinted from Ganan Y, et al. 1998. *Dev Biol* **196:** 33–41, ©1998 with permission from Elsevier. **10.10, left two panels**, Reprinted from Lindsten T, et al. 2000. *Mol Cell* **6:** 1389–1399, ©2000 with permission from Elsevier; **right two panels**, Reprinted from Hubner A, et al. 2010. *Mol Cell Biol* **30:** 98–105, ©2010 with permission from American Society for Microbiology. **10.11**, Adapted from Coucouvanis E, Martin G. 1995. *Cell* **83:** 279–287, ©1995 with permission from Elsevier. **10.12**, Reprinted from Kerr JFR, et al. 1974. *J Cell Sci* **14:** 571–585, ©1974 with permission from The Company of Biologists.

Chapter 11: 11.3 and **11.4**, Modified from Pelengaris S, et al. 2002. *Cell* **109:** 321–334, ©2002 with permission from Elsevier. **11.5**, Modified from Debnath J, et al. 2002. *Cell* **111:** 29–40, ©2002 with permission from Elsevier.

Chapter 12: 12.1, Redrawn with permission from Macmillan Publishers Ltd.: Spencer SL, et al. 2009. *Nature* **459:** 428–432, ©2009. **12.9**, Reprinted from the American Society of Hematology: Straathof KC, et al. 2005. *Blood* **105:** 4247–4254, ©2005. **12.10, left**, PDB 2BZW (Lee K-H, Han W-D, Kim K-J, Oh B-H. The crystal structure of Bcl-xL in complex with full-length BAD; To Be Published); **right**, PDB 2YXJ (Lee EF, Czabotar PE, Smith BJ, Deshayes K, Zobel K, Colman PM, Fairlie WD. 2007. *Cell Death Differ* **14:** 1711–1713). **12.11**, PDB 1RV1 (Vassilev LT, Vu BT, Graves B, Carvajal D, Podlaski F, Filipovic Z, Kong N, Kammlott U, Lukacs C, Klein C, et al. 2004. *Science* **303:** 844–848). **12.13**, Reprinted from He, et al. 2009. *Cell* **137:** 1100–1111, ©2009 with permission from Elsevier.

Additional Reading

CHAPTER 1

Cell Death Nomenclature

Andre N. 2003. Hippocrates of Cos and apoptosis. *Lancet* **361:** 1306.
  A note on an early use of "apoptosis" in medicine.

Degli-Esposti M. 1998. Apoptosis: Who was first? *Cell Death Differ* **5:** 719.
  A hunt for the origins of the word "apoptosis."

Kroemer G, Galluzzi L, Vandenabeele P, Abrams J, Alnemri ES, Baehrecke EH, Blagosklonny MV, El-Deiry WS, Golstein P, Green DR, et al. 2009. Classification of cell death: Recommendations of the Nomenclature Committee on Cell Death 2009. *Cell Death Differ* **16:** 3–11.
  More than simply nomenclature, this review outlines the different types of cell death and problems with their classification.

Lockshin RA, Zakeri Z. 2004. Apoptosis, autophagy, and more. *Int J Biochem Cell Biol* **36:** 2405–2419.
  A useful review of the types of cell death.

History of Apoptosis

Diamantis A, Magiorkinis E, Sakorafas GH, Androutsos G. 2008. A brief history of apoptosis: From ancient to modern times. *Onkologie* **31:** 702–706.
  Not essential reading, but an interesting overview of the study of apoptosis.

Horvitz HR. 2003. Nobel lecture. Worms, life and death. *Biosci Rep* **23:** 239–303.
  The modern study of apoptosis was ignited by the identification of genes that control cell death during development in the nematode C. elegans. This is the Nobel lecture that describes those studies.

Kerr JF, Wyllie AH, Currie AR. 1972. Apoptosis: A basic biological phenomenon with wide-ranging implications in tissue kinetics. *Br J Cancer* **26:** 239–257.
  This is the original paper formally describing apoptosis.

Saunders JW Jr. 1966. Death in embryonic systems. *Science* **154:** 604–612.
  A classic early overview of the role of cell death in development.

Wyllie AH, Kerr JF, Currie AR. 1980. Cell death: The significance of apoptosis. *Int Rev Cytol* **68:** 251–306.
An early review that details the state of knowledge on apoptosis before molecular mechanisms became understood.

CHAPTER 2

Caspases

Lamkanfi M, Festjens N, Declercq W, Vanden Berghe T, Vandenabeele P. 2007. Caspases in cell survival, proliferation and differentiation. *Cell Death Differ* **14:** 44–55.
A review of the caspases and their functions in life and death.

Yuan J, Shaham S, Ledoux S, Ellis HM, Horvitz HR. 1993. The *C. elegans* cell death gene *ced-3* encodes a protein similar to mammalian interleukin-1 β-converting enzyme. *Cell* **75:** 641–652.
The original landmark paper that identified caspases as being important in apoptosis.

Fernandes-Alnemri T, Litwack G, Alnemri ES. 1994. CPP32, a novel human apoptotic protein with homology to *Caenorhabditis elegans* cell death protein Ced-3 and mammalian interleukin-1 β-converting enzyme. *J Biol Chem* **269:** 30761–30764.
The first description of mammalian caspase-3, originally called CPP32.

Alnemri ES, Livingston DJ, Nicholson DW, Salvesen G, Thornberry NA, Wong WW, Yuan J. 1996. Human ICE/CED-3 protease nomenclature. *Cell* **87:** 171.

Eckhart L, Ballaun C, Hermann M, VandeBerg JL, Sipos W, Uthman A, Fischer H. Tschachler E. 2008. Identification of novel mammalian caspases reveals an important role of gene loss in shaping the human caspase repertoire. *Mol Biol Evol* **25:** 831–841.
An overview of the evolutionary relationships among mammalian caspases, including caspase-15, -16, -17, and -18.

Kuida K, Zheng TS, Na S, Kuan C, Yang D, Karasuyama H, Rakic P, Flavell RA. 1996. Decreased apoptosis in the brain and premature lethality in CPP32-deficient mice. *Nature* **384:** 368–372.
The first knockout of caspase-3 (called CPP32) and effects on development.

Caspase Specificities

Timmer JC, Salvesen GS. 2007. Caspase substrates. *Cell Death Differ* **14:** 66–72.

Thornberry NA, Rano TA, Peterson EP, Rasper DM, Timkey T, Garcia-Calvo M, Houtzager VM, Nordstrom PA, Roy S, Vaillancourt JP, et al. 1997. A combinatorial approach defines specificities of members of the caspase family and granzyme B. Functional relationships established for key mediators of apoptosis. *J Biol Chem* **272:** 17907–17911.
The identification of caspase preferences through the use of a combinatorial peptide library.

Mahrus S, Trinidad JC, Barkan DT, Sali A, Burlingame AL, Wells JA. 2008. Global sequencing of proteolytic cleavage sites in apoptosis by specific labeling of protein N termini. *Cell* **134:** 866–876.
Another approach to characterization of caspase substrates.

Functional Caspase Substrates

Luthi AU, Martin SJ. 2007. The CASBAH: A searchable database of caspase substrates. *Cell Death Differ* **14:** 641–650.
An introduction to a searchable database of all known caspase substrates.

Sakahira H, Enari M, Nagata S. 1998. Cleavage of CAD inhibitor in CAD activation and DNA degradation during apoptosis. *Nature* **391:** 96–99.
   One of the original descriptions of the iCAD–CAD system and its role in DNA fragmentation during apoptosis.

Zhang J, Liu X, Scherer DC, van Kaer L, Wang X, Xu M. 1998. Resistance to DNA fragmentation and chromatin condensation in mice lacking the DNA fragmentation factor 45. *Proc Natl Acad Sci* **95:** 12480–12485.
   A clear demonstration of the role of iCAD/DFF45 in DNA fragmentation versus cell death.

Sebbagh M, Renvoize C, Hamelin J, Riche N, Bertoglio J, Breard J. 2001. Caspase-3-mediated cleavage of ROCK I induces MLC phosphorylation and apoptotic membrane blebbing. *Nat Cell Biol* **3:** 346–352.
   One of the original descriptions of ROCK as a caspase substrate and its role in blebbing during apoptosis.

Coleman ML, Sahai EA, Yeo M, Bosch M, Dewar A, Olson MF. 2001. Membrane blebbing during apoptosis results from caspase-mediated activation of ROCK I. *Nat Cell Biol* **3:** 339–345.
   See above.

Kothakota S, Azuma T, Reinhard C, Klippel A, Tang A, Chu K, McGarry TJ, Kirschner MW, Koths K, Kwiatkowski DJ, et al. 1997. Caspase-3-generated fragment of gelsolin: Effector of morphological change in apoptosis. *Science* **278:** 294–298.
   The role of gelsolin in blebbing during apoptosis.

Ricci JE, Munoz-Pinedo C, Fitzgerald P, Bailly-Maitre B, Perkins GA, Yadava N, Scheffler IE, Ellisman MH, Green DR. 2004. Disruption of mitochondrial function during apoptosis is mediated by caspase cleavage of the p75 subunit of complex I of the electron transport chain. *Cell* **117:** 773–786.
   The original description of NDUFS1 as a mitochondrial caspase substrate.

CHAPTER 3

Mechanisms of Caspase Activation

Fuentes-Prior P, Salvesen GS. 2004. The protein structures that shape caspase activity, specificity, activation and inhibition. *Biochem J* **384:** 201–232.

Pop C, Salvesen GS. 2009. Human caspases: Activation, specificity, and regulation. *J Biol Chem* **284:** 21777–21781.

Salvesen GS, Riedl SJ. 2008. Caspase mechanisms. *Adv Exp Med Biol* **615:** 13–23.

Park HH, Lo YC, Lin SC, Wang L, Yang JK, Wu H. 2007. The death domain superfamily in intracellular signaling of apoptosis and inflammation. *Annu Rev Immunol* **25:** 561–586.
   An extensive overview of the structures involved in caspase activation. Several of the mechanisms discussed are covered in later chapters.

Induced Proximity

Muzio M, Stockwell BR, Stennicke HR, Salvesen GS, Dixit VM. 1998. An induced proximity model for caspase-8 activation. *J Biol Chem* **273:** 2926–2930.
   The original description of induced proximity, applied to caspase-8.

Stennicke HR, Deveraux QL, Humke EW, Reed JC, Dixit VM, Salvesen GS. 1999. Caspase-9 can be activated without proteolytic processing. *J Biol Chem* **274:** 8359–8362.

Boatright KM, Renatus M, Scott FL, Sperandio S, Shin H, Pedersen IM, Ricci JE, Edris WA, Sutherlin DP, Green DR, et al. 2003. A unified model for apical caspase activation. *Mol Cell* **11:** 529–541.

Caspase Activation Cascades

Fernandes-Alnemri T, Armstrong RC, Krebs J, Srinivasula SM, Wang L, Bullrich F, Fritz LC, Trapani JA, Tomaselli KJ, Litwack G, et al. 1996. In vitro activation of CPP32 and Mch3 by Mch4, a novel human apoptotic cysteine protease containing two FADD-like domains. *Proc Natl Acad Sci* **93:** 7464–7469.
  One of the first papers to show a caspase activation cascade, with caspase-8 (called Mch4) cleaving and activating caspase-3 (called CPP32) and caspase-7 (called Mch3).

Slee EA, Adrain C, Martin SJ. 1999. Serial killers: Ordering caspase activation events in apoptosis. *Cell Death Differ* **6:** 1067–1074.

Slee EA, Harte MT, Kluck RM, Wolf BB, Casiano CA, Newmeyer DD, Wang HG, Reed JC, Nicholson DW, Alnemri ES, et al. 1999. Ordering the cytochrome c-initiated caspase cascade: Hierarchical activation of caspases-2, -3, -6, -7, -8, and -10 in a caspase-9-dependent manner. *J Cell Biol* **144:** 281–292.
  The order of caspase cleavage following activation of caspase-9 in a cell extract (see Chapter 4 for the initial activation mechanism).

IAPs

Salvesen GS, Riedl SJ. 2007. Caspase inhibition, specifically. *Structure* **15:** 513–514.
  An overview of IAPs and other intracellular caspase inhibitors.

Silke J, Vaux DL. 2001. Two kinds of BIR-containing protein: Inhibitors of apoptosis, or required for mitosis. *J Cell Sci* **114:** 1821–1827.
  An important distinction between different IAPs and their functions.

Vaux DL, Silke J. 2005. IAPs, RINGs and ubiquitylation. *Nat Rev Mol Cell Biol* **6:** 287–297.

Orme M, Meier P. 2009. Inhibitor of apoptosis proteins in *Drosophila*: Gatekeepers of death. *Apoptosis* **14:** 950–960.

Hay BA, Wassarman DA, Rubin GM. 1995. *Drosophila* homologs of baculovirus inhibitor of apoptosis proteins function to block cell death. *Cell* **83:** 1253–1262.
  The discovery of IAPs.

Deveraux QL, Takahashi R, Salvesen GS, Reed JC. 1997. X-linked IAP is a direct inhibitor of cell-death proteases. *Nature* **388:** 300–304.
  The first paper to show that XIAP is a caspase inhibitor.

Riedl SJ, Renatus M, Schwarzenbacher R, Zhou Q, Sun C, Fesik SW, Liddington RC, Salvesen GS. 2001. Structural basis for the inhibition of caspase-3 by XIAP. *Cell* **104:** 791–800.

CHAPTER 4

Mitochondria and Cell Death

Tait SWG, Green DR. 2010. Mitochondria and cell death: Outer membrane permeabilization and beyond. *Nat Rev Cell Mol Biol* **11:** 621–632.
  A detailed review of MOMP and its consequences.

Green DR, Kroemer G. 2004. The pathophysiology of mitochondrial cell death. *Science* **305:** 626–629.

Wallace DC, Fan W. 2009. The pathophysiology of mitochondrial disease as modeled in the mouse. *Genes Dev* **23:** 1714–1736.
 This review discusses how defects in mitochondrial function affect cell survival and death.

Cytochrome c and the Apoptosome

Green DR. 1998. Apoptotic pathways: The roads to ruin. *Cell* **94:** 695–698.
 An early review describing the mitochondrial pathway of apoptosis.

Green DR. 2005. Apoptotic pathways: Ten minutes to dead. *Cell* **121:** 671–674.
 A later review that incorporates MOMP and additional players into the mitochondrial pathway of apoptosis.

Liu X, Kim CN, Yang J, Jemmerson R, Wang X. 1996. Induction of apoptotic program in cell-free extracts: Requirement for dATP and cytochrome c. *Cell* **86:** 147–157.
 The original paper showing that cytochrome c induces caspase activation.

Zou H, Henzel WJ, Liu X, Lutschg A, Wang X. 1997. Apaf-1, a human protein homologous to *C. elegans* CED-4, participates in cytochrome c-dependent activation of caspase-3. *Cell* **90:** 405–413.
 The identification and characterization of APAF1.

Li P, Nijhawan D, Budihardjo I, Srinivasula SM, Ahmad M, Alnemri ES, Wang X. 1997. Cytochrome c and dATP-dependent formation of Apaf-1/caspase-9 complex initiates an apoptotic protease cascade. *Cell* **91:** 479–489.
 The interaction between APAF1 and caspase-9 leading to caspase activation was first described in this paper.

Yuan S, Yu X, Topf M, Ludtke SJ, Wang X, Akey CW. 2010. Structure of an apoptosome-procaspase-9 CARD complex. *Structure* **18:** 571–583.

MOMP

Goldstein JC, Waterhouse NJ, Juin P, Evan GI, Green DR. 2000. The coordinate release of cytochrome c during apoptosis is rapid, complete and kinetically invariant. *Nat Cell Biol* **2:** 156–162.
 The original paper demonstrating MOMP in cells using fluorescent cytochrome c.

Waterhouse NJ, Goldstein JC, von Ahsen O, Schuler M, Newmeyer DD, Green DR. 2001. Cytochrome c maintains mitochondrial transmembrane potential and ATP generation after outer mitochondrial membrane permeabilization during the apoptotic process. *J Cell Biol* **153:** 319–328.
 The immediate consequences of MOMP for mitochondria are explored.

Smac/DIABLO and Omi

Verhagen AM, Kratina TK, Hawkins CJ, Silke J, Ekert PG, Vaux DL. 2007. Identification of mammalian mitochondrial proteins that interact with IAPs via N-terminal IAP binding motifs. *Cell Death Differ* **14:** 348–357.

Green DR. 2000. Apoptotic pathways: Paper wraps stone blunts scissors. *Cell* **102:** 1–4.

Du C, Fang M, Li Y, Li L, Wang X. 2000. Smac, a mitochondrial protein that promotes cytochrome c-dependent caspase activation by eliminating IAP inhibition. *Cell* **102:** 33–42.
 The original description of Smac, published with the original description of the mouse homolog DIABLO.

Verhagen AM, Ekert PG, Pakusch M, Silke J, Connolly LM, Reid GE, Moritz RL, Simpson RJ, Vaux DL. 2000. Identification of DIABLO, a mammalian protein that promotes apoptosis by binding to and antagonizing IAP proteins. *Cell* **102:** 43–53.
   See above.

Hegde R, Srinivasula SM, Zhang Z, Wassell R, Mukattash R, Cilenti L, DuBois G, Lazebnik Y, Zervos AS, Fernandes-Alnemri T, et al. 2002. Identification of Omi/HtrA2 as a mitochondrial apoptotic serine protease that disrupts inhibitor of apoptosis protein–caspase interaction. *J Biol Chem* **277:** 432–438.
   One of several papers initially describing Omi and its role in apoptosis.

Caspase-independent Cell Death

Chipuk JE, Green DR. 2005. Do inducers of apoptosis trigger caspase-independent cell death? *Nat Rev Mol Cell Biol* **6:** 268–275.

Lartigue L, Kushnareva Y, Seong Y, Lin H, Faustin B, Newmeyer DD. 2009. Caspase-independent mitochondrial cell death results from loss of respiration, not cytotoxic protein release. *Mol Biol Cell* **20:** 4871–4884.
   A study showing that MOMP-induced, caspase-independent cell death is a consequence of loss of mitochonrial function.

Susin SA, Lorenzo HK, Zamzami N, Marzo I, Snow BE, Brothers GM, Mangion J, Jacotot E, Costantini P, Loeffler M, et al. 1999. Molecular characterization of mitochondrial apoptosis-inducing factor. *Nature* **397:** 441–446.
   The initial characterization of AIF.

Li LY, Luo X, Wang X. 2001. Endonuclease G is an apoptotic DNase when released from mitochondria. *Nature* **412:** 95–99.
   Initial description of endonuclease G and its proposed role in caspase-independent cell death.

Mitochondrial Permeability Transition

Lemasters JJ, Theruvath TP, Zhong Z, Nieminen AL. 2009. Mitochondrial calcium and the permeability transition in cell death. *Biochim Biophys Acta* **1787:** 1395–1401.
   An overview of the permeability transition in cell death, but one that suggests roles in apoptosis.

Kroemer G, Galluzzi L, Brenner C. 2007. Mitochondrial membrane permeabilization in cell death. *Physiol Rev* **87:** 99–163.
   An attempt to unify concepts of MOMP and MPT in cell death.

Tsujimoto Y, Nakagawa T, Shimizu S. 2006. Mitochondrial membrane permeability transition and cell death. *Biochim Biophys Acta* **1757:** 1297–1300.

Leung AW, Halestrap AP. 2008. Recent progress in elucidating the molecular mechanism of the mitochondrial permeability transition pore. *Biochim Biophys Acta* **1777:** 946–952.

Nakagawa T, Shimizu S, Watanabe T, Yamaguchi O, Otsu K, Yamagata H, Inohara H, Kubo T, Tsujimoto Y. 2005. Cyclophilin D-dependent mitochondrial permeability transition regulates some necrotic but not apoptotic cell death. *Nature* **434:** 652–658.
   One of several papers showing that MPT has roles in necrosis but not apoptosis.

Apoptosomes in Worms and Flies

Bao Q, Shi Y. 2007. Apoptosome: A platform for the activation of initiator caspases. *Cell Death Differ* **14:** 56–65.

Qi S, Pang Y, Hu Q, Liu Q, Li H, Zhou Y, He T, Liang Q, Liu Y, Yuan X, et al. 2010. Crystal structure of the *Caenorhabditis elegans* apoptosome reveals an octameric assembly of CED-4. *Cell* **141:** 446–457.

Yu X, Wang L, Acehan D, Wang X, Akey CW. 2006. Three-dimensional structure of a double apoptosome formed by the *Drosophila* Apaf-1 related killer. *J Mol Biol* **355:** 577–589.

Rodriguez A, Chen P, Oliver H, Abrams JM. 2002. Unrestrained caspase-dependent cell death caused by loss of Diap1 function requires the *Drosophila* Apaf-1 homolog, Dark. *EMBO J* **21:** 2189–2197.
 The role of DIAP1 in restraining ARK (called "Dark")-induced caspase activity in flies.

Rodriguez A, Oliver H, Zou H, Chen P, Wang X, Abrams JM. 1999. Dark is a *Drosophila* homologue of Apaf-1/CED-4 and functions in an evolutionarily conserved death pathway. *Nat Cell Biol* **1:** 272–279.
 The identification of ARK (called "Dark").

Yuan J, Horvitz HR. 1992. The *Caenorhabditis elegans* cell death gene *ced-4* encodes a novel protein and is expressed during the period of extensive programmed cell death. *Development* **116:** 309–320.
 The original characterization of CED4.

Chinnaiyan AM, Chaudhary D, O'Rourke K, Koonin EV, Dixit VM. 1997. Role of CED-4 in the activation of CED-3. *Nature* **388:** 728–729.
 One of the original papers showing that CED4 biochemically activates CED3.

CHAPTER 5

BCL-2 Proteins

Chipuk JE, Moldoveanu T, Llambi F, Parsons MJ, Green DR. 2010. The BCL-2 family reunion. *Mol Cell* **37:** 299–310.

Youle RJ, Strasser A. 2008. The BCL-2 protein family: Opposing activities that mediate cell death. *Nat Rev Mol Cell Biol* **9:** 47–59.

Anti-apoptotic BCL-2 Proteins

Vaux DL, Cory S, Adams JM. 1988. *Bcl-2* gene promotes haemopoietic cell survival and cooperates with c-myc to immortalize pre-B cells. *Nature* **335:** 440–442.
 The first paper to show that BCL-2 preserves cell survival in primary and transformed cells.

Hockenbery D, Nunez G, Milliman C, Schreiber RD, Korsmeyer SJ. 1990. Bcl-2 is an inner mitochondrial membrane protein that blocks programmed cell death. *Nature* **348:** 334–336.
 The first paper to show that BCL-2 is associated with mitochondria (the original assertion that it is on the inner membrane was due to an artifact) and that BCL-2 inhibits apoptosis.

Kluck RM, Bossy-Wetzel E, Green DR, Newmeyer DD. 1997. The release of cytochrome c from mitochondria: A primary site for Bcl-2 regulation of apoptosis. *Science* **275:** 1132–1136.
 One of two papers (see below) identifying the role of BCL-2 in blocking the release of cytochrome c from mitochondria.

Yang J, Liu X, Bhalla K, Kim CN, Ibrado AM, Cai J, Peng TI, Jones DP, Wang X. 1997. Prevention of apoptosis by Bcl-2: Release of cytochrome c from mitochondria blocked. *Science* **275:** 1129–1132.
 See above.

Vaux DL, Weissman IL, Kim SK. 1992. Prevention of programmed cell death in *Caenorhabditis elegans* by human bcl-2. *Science* **258:** 1955–1957.
 One of two papers that showed that CED9 and BCL-2 are functionally equivalent in nematodes. Although BCL-2 does not bind to CED4, it can sequester EGL1, which was not shown until later.

Hengartner MO, Horvitz HR. 1994. *C. elegans* cell survival gene *ced-9* encodes a functional homolog of the mammalian proto-oncogene *bcl-2*. *Cell* **76:** 665–676.
   See above.

Jabbour AM, Puryer MA, Yu JY, Lithgow T, Riffkin CD, Ashley DM, Vaux DL, Ekert PG, Hawkins CJ. 2006. Human Bcl-2 cannot directly inhibit the *Caenorhabditis elegans* Apaf-1 homologue CED-4, but can interact with EGL-1. *J Cell Sci* **119:** 2572–2582.
   See above.

Pro-apoptotic BCL-2 Effectors

Oltvai ZN, Milliman CL, Korsmeyer SJ. 1993. Bcl-2 heterodimerizes in vivo with a conserved homolog, Bax, that accelerates programmed cell death. *Cell* **74:** 609–619.
   The discovery of BAX.

Chittenden T, Harrington EA, O'Connor R, Flemington C, Lutz RJ, Evan GI, Guild BC. 1995. Induction of apoptosis by the Bcl-2 homologue Bak. *Nature* **374:** 733–736.
   The discovery of BAK.

Wei MC, Zong WX, Cheng EH, Lindsten T, Panoutsakopoulou V, Ross AJ, Roth KA, MacGregor GR, Thompson CB, Korsmeyer SJ. 2001. Proapoptotic BAX and BAK: A requisite gateway to mitochondrial dysfunction and death. *Science* **292:** 727–730.
   This paper showed that BAX and BAK are necessary for MOMP and apoptosis via the mitochondrial pathway.

BH3-only Proteins

Giam M, Huang DC, Bouillet P. 2008. BH3-only proteins and their roles in programmed cell death. *Oncogene* (suppl 1) **27:** S128–S136.

Stoka V, Turk B, Schendel SL, Kim TH, Cirman T, Snipas SJ, Ellerby LM, Bredesen D, Freeze H, Abrahamson M, et al. 2001. Lysosomal protease pathways to apoptosis. Cleavage of bid, not pro-caspases, is the most likely route. *J Biol Chem* **276:** 3149–3157.
   BID as a protease sensor for engaging the mitochondrial pathway of apoptosis.

Datta SR, Ranger AM, Lin MZ, Sturgill JF, Ma YC, Cowan CW, Dikkes P, Korsmeyer SJ, Greenberg ME. 2002. Survival factor-mediated BAD phosphorylation raises the mitochondrial threshold for apoptosis. *Dev Cell* **3:** 631–643.
   The regulation of BAD by survival factor signaling.

Bouillet P, Metcalf D, Huang DC, Tarlinton DM, Kay TW, Kontgen F, Adams JM, Strasser A. 1999. Proapoptotic Bcl-2 relative Bim required for certain apoptotic responses, leukocyte homeostasis, and to preclude autoimmunity. *Science* **286:** 1735–1738.
   The BIM knockout mouse illustrates roles for this protein in apoptosis.

Puthalakath H, Huang DC, O'Reilly LA, King SM, Strasser A. 1999. The proapoptotic activity of the Bcl-2 family member Bim is regulated by interaction with the dynein motor complex. *Mol Cell* **3:** 287–296.
   The association of BIM with the cytoskeleton.

Conradt B, Horvitz HR. 1998. The *C. elegans* protein EGL-1 is required for programmed cell death and interacts with the Bcl-2-like protein CED-9. *Cell* **93:** 519–529.
   EGL1 is a BH3-only protein in nematodes.

BCL-2 Protein Interactions

Letai A, Bassik MC, Walensky LD, Sorcinelli MD, Weiler S, Korsmeyer SJ. 2002. Distinct BH3 domains either sensitize or activate mitochondrial apoptosis, serving as prototype cancer therapeutics. *Cancer Cell* **2:** 183–192.
  The proposal of the "direct activator/derepressor" (called "sensitizer") model of BH3-only protein function.

Chen L, Willis SN, Wei A, Smith BJ, Fletcher JI, Hinds MG, Colman PM, Day CL, Adams JM, Huang DC. 2005. Differential targeting of prosurvival Bcl-2 proteins by their BH3-only ligands allows complementary apoptotic function. *Mol Cell* **17:** 393–403.
  The "neutralization" model of BH3-only protein function and specificities of BH3-only proteins for different anti-apoptotic proteins.

Willis SN, Chen L, Dewson G, Wei A, Naik E, Fletcher JI, Adams JM, Huang DC. 2005. Proapoptotic Bak is sequestered by Mcl-1 and Bcl-xL, but not Bcl-2, until displaced by BH3-only proteins. *Genes Dev* **19:** 1294–1305.
  The "neutralization" model of BH3-only protein function.

Kuwana T, Mackey MR, Perkins G, Ellisman MH, Latterich M, Schneiter R, Green DR, Newmeyer DD. 2002. Bid, Bax, and lipids cooperate to form supramolecular openings in the outer mitochondrial membrane. *Cell* **111:** 331–342.
  Biochemical characterization of BAX activation and membrane permeabilization.

Kim H, Tu HC, Ren D, Takeuchi O, Jeffers JR, Zambetti GP, Hsieh JJ, Cheng EH. 2009. Stepwise activation of BAX and BAK by tBID, BIM, and PUMA initiates mitochondrial apoptosis. *Mol Cell* **36:** 487–499.
  Further analysis of the "direct activator/derepressor" model of BH3-only protein function.

Lovell JF, Billen LP, Bindner S, Shamas-Din A, Fradin C, Leber B, Andrews DW. 2008. Membrane binding by tBid initiates an ordered series of events culminating in membrane permeabilization by Bax. *Cell* **135:** 1074–1084.
  Use of FRET technology to characterize derepression and direct activation of BAX.

Gavathiotis E, Suzuki M, Davis ML, Pitter K, Bird GH, Katz SG, Tu HC, Kim H, Cheng EH, Tjandra N, et al. 2008. BAX activation is initiated at a novel interaction site. *Nature* **455:** 1076–1081.
  Structural analysis of the binding of the BIM BH3 region to BAX and early steps in BAX activation.

Other Functions of the BCL-2 Family

Karbowski M, Lee YJ, Gaume B, Jeong SY, Frank S, Nechushtan A, Santel A, Fuller M, Smith CL, Youle RJ. 2002. Spatial and temporal association of Bax with mitochondrial fission sites, Drp1, and Mfn2 during apoptosis. *J Cell Biol* **159:** 931–938.
  Characterization of interactions between BAX and proteins involved in mitochondrial fission and fusion.

Sheridan C, Delivani P, Cullen SP, Martin SJ. 2008. Bax- or Bak-induced mitochondrial fission can be uncoupled from cytochrome c release. *Mol Cell* **31:** 570–585.
  Evidence that mitochondrial fission occurs at the time of MOMP, but is not required for permeabilization.

Rong Y, Distelhorst CW. 2008. Bcl-2 protein family members: Versatile regulators of calcium signaling in cell survival and apoptosis. *Annu Rev Physiol* **70:** 73–91.

Chen R, Valencia I, Zhong F, McColl KS, Roderick HL, Bootman MD, Berridge MJ, Conway SJ, Holmes AB, Mignery GA, et al. 2004. Bcl-2 functionally interacts with inositol 1,4,5-trisphosphate receptors

to regulate calcium release from the ER in response to inositol 1,4,5-trisphosphate. *J Cell Biol* **166:** 193–203.
  Characterization of the regulation of calcium homeostasis by BCL-2.

Levine B, Sinha S, Kroemer G. 2008. Bcl-2 family members: Dual regulators of apoptosis and autophagy. *Autophagy* **4:** 600–606.

Pattingre S, Tassa A, Qu X, Garuti R, Liang XH, Mizushima N, Packer M, Schneider MD, Levine B. 2005. Bcl-2 antiapoptotic proteins inhibit Beclin 1-dependent autophagy. *Cell* **122:** 927–939.
  A mechanism for the regulation of autophagy by BCL-2. Autophagy is covered in more detail in Chapter 8.

Zhang H, Bosch-Marce M, Shimoda LA, Tan YS, Baek JH, Wesley JB, Gonzalez FJ, Semenza GL. 2008. Mitochondrial autophagy is an HIF-1-dependent adaptive metabolic response to hypoxia. *J Biol Chem* **283:** 10892–10903.
  How BCL-2 interactions can cause autophagic removal of mitochondria.

Cuconati A, White E. 2002. Viral homologs of BCL-2: Role of apoptosis in the regulation of virus infection. *Genes Dev* **16:** 2465–2478.

CHAPTER 6

Death Receptors

Wilson NS, Dixit V, Ashkenazi A. 2009. Death receptor signal transducers: Nodes of coordination in immune signaling networks. *Nat Immunol* **10:** 348–355.

Guicciardi ME, Gores GJ. 2009. Life and death by death receptors. *FASEB J* **23:** 1625–1637.

Falschlehner C, Ganten TM, Koschny R, Schaefer U, Walczak H. 2009. TRAIL and other TRAIL receptor agonists as novel cancer therapeutics. *Adv Exp Med Biol* **647:** 195–206.

Wallach D, Varfolomeev EE, Malinin NL, Goltsev YV, Kovalenko AV, Boldin MP. 1999. Tumor necrosis factor receptor and Fas signaling mechanisms. *Annu Rev Immunol* **17:** 331–367.

Banner DW, D'Arcy A, Janes W, Gentz R, Schoenfeld HJ, Broger C, Loetscher H, Lesslauer W. 1993. Crystal structure of the soluble human 55 kd TNF receptor-human TNF β complex: Implications for TNF receptor activation. *Cell* **73:** 431–445.

Hymowitz SG, Christinger HW, Fuh G, Ultsch M, O'Connell M, Kelley RF, Ashkenazi A, de Vos AM. 1999. Triggering cell death: The crystal structure of Apo2L/TRAIL in a complex with death receptor 5. *Mol Cell* **4:** 563–571.
  The structure of TRAIL shows a role for zinc that is not seen in other TNF-family ligands.

Death Receptor Signaling

Muzio M, Stockwell BR, Stennicke HR, Salvesen GS, Dixit VM. 1998. An induced proximity model for caspase-8 activation. *J Biol Chem* **273:** 2926–2930.
  The activation of caspase-8 by FADD-mediated induced proximity.

Micheau O, Tschopp J. 2003. Induction of TNF receptor I-mediated apoptosis via two sequential signaling complexes. *Cell* **114:** 181–190.
  The identification of two complexes involved in TNFR signaling.

Karin M, Gallagher E. 2009. TNFR signaling: Ubiquitin-conjugated TRAFfic signals control stop-and-go for MAPK signaling complexes. *Immunol Rev* **228:** 225–240.
  An in-depth overview of TNFR signaling to NF-κB and other pathways.

Defects in the CD95 Pathway Cause Disease

Bidere N, Su HC, Lenardo MJ. 2006. Genetic disorders of programmed cell death in the immune system. *Annu Rev Immunol* **24:** 321–352.
 An overview of diseases in mice and humans caused by defects in CD95, CD95L, and associated caspases.

Watanabe-Fukunaga R, Brannan CI, Copeland NG, Jenkins NA, Nagata S. 1992. Lymphoproliferation disorder in mice explained by defects in Fas antigen that mediates apoptosis. *Nature* **356:** 314–317.
 Defects in CD95 (called "Fas antigen") are responsible for a lymphoaccumulative disease in mice (originally thought to be a lymphoproliferation defect).

Lynch DH, Watson ML, Alderson MR, Baum PR, Miller RE, Tough T, Gibson M, Davis-Smith T, Smith CA, Hunter K, et al. 1994. The mouse Fas-ligand gene is mutated in gld mice and is part of a TNF family gene cluster. *Immunity* **1:** 131–136.
 Defects in CD95L (called "Fas ligand") are responsible for a lymphoaccumulative disease in mice.

Death Receptors and the Mitochondrial Pathway

Li H, Zhu H, Xu CJ, Yuan J. 1998. Cleavage of BID by caspase 8 mediates the mitochondrial damage in the Fas pathway of apoptosis. *Cell* **94:** 491–501.
 Two papers identified BID as a target for caspase-8, linking the death receptor and mitochondrial pathways of apoptosis.

Luo X, Budihardjo I, Zou H, Slaughter C, Wang X. 1998. Bid, a Bcl2 interacting protein, mediates cytochrome c release from mitochondria in response to activation of cell surface death receptors. *Cell* **94:** 481–490.
 See above.

Scaffidi C, Fulda S, Srinivasan A, Friesen C, Li F, Tomaselli KJ, Debatin KM, Krammer PH, Peter ME. 1998. Two CD95 (APO-1/Fas) signaling pathways. *EMBO J* **17:** 1675–1687.
 The description of two cell types ("type I and type II") in which the mitochondrial pathway is dispensable (type I) or required (type II) for death receptor–induced apoptosis.

Jost PJ, Grabow S, Gray D, McKenzie MD, Nachbur U, Huang DC, Bouillet P, Thomas HE, Borner C, Silke J, et al. 2009. XIAP discriminates between type I and type II FAS-induced apoptosis. *Nature* **460:** 1035–1039.
 Inhibition of XIAP is required for apoptosis induced by CD95 (called "Fas") in type II cells in mice.

CHAPTER 7

The Inflammatory Caspases

Kersse K, Vanden Berghe T, Lamkanfi M, Vandenabeele P. 2007. A phylogenetic and functional overview of inflammatory caspases and caspase-1-related CARD-only proteins. *Biochem Soc Trans* **35:** 1508–1511.
 An overview of the inflammatory caspases and the molecules that interact with them.

Cerretti DP, Kozlosky CJ, Mosley B, Nelson N, Van Ness K, Greenstreet TA, March CJ, Kronheim SR, Druck T, Cannizzaro LA, et al. 1992. Molecular cloning of the interleukin-1 β converting enzyme. *Science* **256:** 97–100.
 Two papers describing the identification of caspase-1, the first caspase discovered.

Thornberry NA, Bull HG, Calaycay JR, Chapman KT, Howard AD, Kostura MJ, Miller DK, Molineaux SM, Weidner JR, Aunins J, et al. 1992. A novel heterodimeric cysteine protease is required for interleukin-1 β processing in monocytes. *Nature* **356:** 768–774.
   See above.

Dinarello CA. 2009. Immunological and inflammatory functions of the interleukin-1 family. *Annu Rev Immunol* **27:** 519–550.
   Interleukin-1 is a major target of caspase-1, and an understanding of its biological roles provides a context for appreciating the significance of caspase-1 function.

Keller M, Ruegg A, Werner S, Beer HD. 2008. Active caspase-1 is a regulator of unconventional protein secretion. *Cell* **132:** 818–831.
   The role of caspase-1 in secretion of proteins that lack conventional secretory signal sequences.

TLRs and NLRs

Fukata M, Vamadevan AS, Abreu MT. 2009. Toll-like receptors (TLRs) and Nod-like receptors (NLRs) in inflammatory disorders. *Semin Immunol* **21:** 242–253.

O'Neill LA. 2008. The interleukin-1 receptor/Toll-like receptor superfamily: 10 years of progress. *Immunol Rev* **226:** 10–18.

Medzhitov R, Preston-Hurlburt P, Janeway CA Jr. 1997. A human homologue of the *Drosophila* Toll protein signals activation of adaptive immunity. *Nature* **388:** 394–397.
   The original paper describing mammalian Toll-like receptors and their importance in the immune response.

Inflammasomes

Schroder K, Tschopp J. 2010. The inflammasomes. *Cell* **140:** 821–832.

Franchi L, Eigenbrod T, Munoz-Planillo R, Nunez G. 2009. The inflammasome: A caspase-1-activation platform that regulates immune responses and disease pathogenesis. *Nat Immunol* **10:** 241–247.

Martinon F, Burns K, Tschopp J. 2002. The inflammasome: A molecular platform triggering activation of inflammatory caspases and processing of proIL- β. *Mol Cell* **10:** 417–426.
   The first paper describing an inflammasome, which in this case involved NLRP1 (called "Nalp1"). Although this inflammasome includes caspase-5, all others described do not.

Jin C, Flavell RA. 2010. Molecular mechanism of NLRP3 inflammasome activation. *J Clin Immunol* (in press).

Agostini L, Martinon F, Burns K, McDermott MF, Hawkins PN, Tschopp J. 2004. NALP3 forms an IL-1β-processing inflammasome with increased activity in Muckle–Wells autoinflammatory disorder. *Immunity* **20:** 319–325.
   A description of the NLRP3 (called "NALP3") inflammasome.

Mariathasan S, Newton K, Monack DM, Vucic D, French DM, Lee WP, Roose-Girma M, Erickson S, Dixit VM. 2004. Differential activation of the inflammasome by caspase-1 adaptors ASC and Ipaf. *Nature* **430:** 213–218.
   A description of the IPAF inflammasome.

Apoptosis via Caspase-1 Activation

Kepp O, Galluzzi L, Zitvogel L, Kroemer G. 2010. Pyroptosis—A cell death modality of its kind? *Eur J Immunol* **40:** 627–630.

Bergsbaken T, Fink SL, Cookson BT. 2009. Pyroptosis: Host cell death and inflammation. *Nat Rev Microbiol* **7:** 99–109.

Fernandes-Alnemri T, Wu J, Yu JW, Datta P, Miller B, Jankowski W, Rosenberg S, Zhang J, Alnemri ES. 2007. The pyroptosome: A supramolecular assembly of ASC dimers mediating inflammatory cell death via caspase-1 activation. *Cell Death Differ* **14:** 1590–1604.
   This paper proposes that caspase-1–mediated cell death is favored by oligomers of ASC, forming independently of other adapter proteins.

Caspase-2

Kumar S. 2009. Caspase 2 in apoptosis, the DNA damage response and tumour suppression: Enigma no more? *Nat Rev Cancer* **9:** 897–903.
   Possible roles for caspase-2 in apoptosis and other phenomena.

Krumschnabel G, Manzl C, Villunger A. 2009. Caspase-2: Killer, savior and safeguard—Emerging versatile roles for an ill-defined caspase. *Oncogene* **28:** 3093–3096.
   Additional views of roles for caspase-2.

Tinel A, Tschopp J. 2004. The PIDDosome, a protein complex implicated in activation of caspase-2 in response to genotoxic stress. *Science* **304:** 843–846.
   The first characterization of a complex responsible for activating caspase-2.

Park HH, Logette E, Raunser S, Cuenin S, Walz T, Tschopp J, Wu H. 2007. Death domain assembly mechanism revealed by crystal structure of the oligomeric PIDDosome core complex. *Cell* **128:** 533–546.

Bouchier-Hayes L, Oberst A, McStay GP, Connell S, Tait SW, Dillon CP, Flanagan JM, Beere HM, Green DR. 2009. Characterization of cytoplasmic caspase-2 activation by induced proximity. *Mol Cell* **35:** 830–840.
   The activation of caspase-2 by several stressors, analyzed using a live cell–imaging technique.

CHAPTER 8

Nonapoptotic Cell Death

Degterev A, Yuan J. 2008. Expansion and evolution of cell death programmes. *Nat Rev Mol Cell Biol* **9:** 378–390.
   See also the reviews listed under Additional Reading for Chapter 1.

Necrosis

Silva MT, do Vale A, dos Santos NM. 2008. Secondary necrosis in multicellular animals: An outcome of apoptosis with pathogenic implications. *Apoptosis* **13:** 463–482.

Bouchard VJ, Rouleau M, Poirier GG. 2003. PARP-1, a determinant of cell survival in response to DNA damage. *Exp Hematol* **31:** 446–454.

Whelan RS, Kaplinskiy V, Kitsis RN. Cell death in the pathogenesis of heart disease: Mechanisms and significance. *Annu Rev Physiol* **72:** 19–44.
   Overview of ischemia/reperfusion injury in the heart and other roles for cell death in heart disease.

Szydlowska K, Tymianski M. 2010. Calcium, ischemia and excitotoxicity. *Cell Calcium* **47:** 122–129.

Kim YS, Morgan MJ, Choksi S, Liu ZG. 2007. TNF-induced activation of the Nox1 NADPH oxidase and its role in the induction of necrotic cell death. *Mol Cell* **26:** 675–687.

Eliasson MJ, Sampei K, Mandir AS, Hurn PD, Traystman RJ, Bao J, Pieper A, Wang ZQ, Dawson TM, Snyder SH, et al. 1997. Poly(ADP-ribose) polymerase gene disruption renders mice resistant to cerebral ischemia. *Nat Med* **3:** 1089–1095.
   A connection between ischemia/reperfusion injury and PARP.

Programmed Necrosis (Necroptosis)

Christofferson DE, Yuan J. 2009. Necroptosis as an alternative form of programmed cell death. *Curr Opin Cell Biol* **22:** 263–268.

Declercq W, Vanden Berghe T, Vandenabeele P. 2009. RIP kinases at the crossroads of cell death and survival. *Cell* **138:** 229–232.

Holler N, Zaru R, Micheau O, Thome M, Attinger A, Valitutti S, Bodmer JL, Schneider P, Seed B, Tschopp J. 2000. Fas triggers an alternative, caspase-8-independent cell death pathway using the kinase RIP as effector molecule. *Nat Immunol* **1:** 489–495.
   The first description of RIPK-dependent cell death, in this case induced by CD95 (called "Fas").

Hitomi J, Christofferson DE, Ng A, Yao J, Degterev A, Xavier RJ, Yuan J. 2008. Identification of a molecular signaling network that regulates a cellular necrotic cell death pathway. *Cell* **135:** 1311–1323.
   An approach to characterization of the mechanism of RIPK-dependent necrosis.

Autophagy

He C, Klionsky DJ. 2009. Regulation mechanisms and signaling pathways of autophagy. *Annu Rev Genet* **43:** 67–93.

Eskelinen EL. 2008. New insights into the mechanisms of macroautophagy in mammalian cells. *Int Rev Cell Mol Biol* **266:** 207–247.

Nakatogawa H, Suzuki K, Kamada Y, Ohsumi Y. 2009. Dynamics and diversity in autophagy mechanisms: Lessons from yeast. *Nat Rev Mol Cell Biol* **10:** 458–467.

Suzuki K, Ohsumi Y. 2010. Current knowledge of the pre-autophagosomal structure (PAS). *FEBS Lett* **584:** 1280–1286.

Autophagic Cell Death

Levine B, Yuan J. 2005. Autophagy in cell death: An innocent convict? *J Clin Invest* **115:** 2679–2688.

Baehrecke EH. 2005. Autophagy: Dual roles in life and death? *Nat Rev Mol Cell Biol* **6:** 505–510.

Yu L, Alva A, Su H, Dutt P, Freundt E, Welsh S, Baehrecke EH, Lenardo MJ. 2004. Regulation of an ATG7-beclin 1 program of autophagic cell death by caspase-8. *Science* **304:** 1500–1502.
   The first paper suggesting that caspase-8 inhibits autophagy-dependent cell death, which ultimately turned out to be RIPK-dependent cell death. The relationships between these modes of cell death remain unresolved.

Mitotic Catastrophe

Vakifahmetoglu H, Olsson M, Zhivotovsky B. 2008. Death through a tragedy: Mitotic catastrophe. *Cell Death Differ* **15:** 1153–1162.

Castedo M, Perfettini JL, Roumier T, Andreau K, Medema R, Kroemer G. 2004. Cell death by mitotic catastrophe: A molecular definition. *Oncogene* **23:** 2825–2837.

Kirsch DG, Santiago PM, di Tomaso E, Sullivan JM, Hou WS, Dayton T, Jeffords LB, Sodha P, Mercer KL, Cohen R, et al. p53 controls radiation-induced gastrointestinal syndrome in mice independent of apoptosis. *Science* **327:** 593–596.
   Cell death in the intestines of BAX–BAK double-deficient mice may be a form of mitotic catastrophe. p53 is discussed in Chapter 11.

CHAPTER 9

Engulfment of Dying Cells

Elliott MR, Ravichandran KS. 2010. Clearance of apoptotic cells: Implications in health and disease. *J Cell Biol* **189:** 1059–1070.

Erwig LP, Henson PM. 2008. Clearance of apoptotic cells by phagocytes. *Cell Death Differ* **15:** 243–250.

Ravichandran KS, Lorenz U. 2007. Engulfment of apoptotic cells: Signals for a good meal. *Nat Rev Immunol* **7:** 964–974.
   A particularly thorough overview of the mechanisms of engulfment.

Lauber K, Blumenthal SG, Waibe MI, Wesselborg S. 2004. Clearance of apoptotic cells: Getting rid of the corpses. *Mol Cell* **14:** 277–287.

Find-me Signals

Lauber K, Bohn E, Krober SM, Xiao YJ, Blumenthal SG, Lindemann RK, Marini P, Wiedig C, Zobywalski A, Baksh S, et al. 2003. Apoptotic cells induce migration of phagocytes via caspase-3-mediated release of a lipid attraction signal. *Cell* **113:** 717–730.
   Identification of lysophosphocholine as a "find-me" signal and one way in which it is produced following caspase activation.

Elliott MR, Chekeni FB, Trampont PC, Lazarowski ER, Kadl A, Walk SF, Park D, Woodson RI, Ostankovich M, Sharma P, et al. 2009. Nucleotides released by apoptotic cells act as a find-me signal to promote phagocytic clearance. *Nature* **461:** 282–286.
   Identification of ATP as a "find-me" signal.

Bind-me Signals

Bevers EM, Williamson PL. 2010. Phospholipid scramblase: An update. *FEBS Lett* **584:** 2724–2730.
   The mechanism of caspase-dependent phosphatidylserine externalization remains unresolved; this describes our state of knowledge.

Martin SJ, Reutelingsperger CP, McGahon AJ, Rader JA, van Schie RC, LaFace DM, Green DR. 1995. Early redistribution of plasma membrane phosphatidylserine is a general feature of apoptosis regardless of the initiating stimulus: Inhibition by overexpression of Bcl-2 and Abl. *J Exp Med* **182:** 1545–1556.
   An early paper describing the use of annexin V to detect apoptosis.

Phagocyte Receptors for Dying Cells

Wu YC, Horvitz HR. 1998. The *C. elegans* cell corpse engulfment gene *ced-7* encodes a protein similar to ABC transporters. *Cell* **93:** 951–960.
   The original characterization of CED7.

Hamon Y, Broccardo C, Chambenoit O, Luciani MF, Toti F, Chaslin S, Freyssinet JM, Devaux PF, McNeish J, Marguet D, et al. 2000. ABC1 promotes engulfment of apoptotic cells and transbilayer redistribution of phosphatidylserine. *Nat Cell Biol* **2:** 399–406.
   ABC1 in mammalian engulfment of dying cells and its possible role in phospholipid scrambling; the relationship to caspases remains unclear.

Scott RS, McMahon EJ, Pop SM, Reap EA, Caricchio R, Cohen PL, Earp HS, Matsushima GK. 2001. Phagocytosis and clearance of apoptotic cells is mediated by MER. *Nature* **411:** 207–211.

Hanayama R, Tanaka M, Miwa K, Shinohara A, Iwamatsu A, Nagata S. 2002. Identification of a factor that links apoptotic cells to phagocytes. *Nature* **417:** 182–187.
   The identification of MFG-E8 as a bridge molecule recognizing phosphatidylserine and its role in engulfment of dying cells.

Miyanishi M, Tada K, Koike M, Uchiyama Y, Kitamura T, Nagata S. 2007. Identification of Tim4 as a phosphatidylserine receptor. *Nature* **450:** 435–439.

Park D, Tosello-Trampont AC, Elliott MR, Lu M, Haney LB, Ma Z, Klibanov AL, Mandell JW, Ravichandran KS. 2007. BAI1 is an engulfment receptor for apoptotic cells upstream of the ELMO/DOCK180/RAC module. *Nature* **450:** 430–434.

Franc NC, Heitzler P, Ezekowitz RA, White K. 1999. Requirement for croquemort in phagocytosis of apoptotic cells in *Drosophila*. *Science* **284:** 1991–1994.

Eat-me Signals and Engulfment

Reddien PW, Horvitz HR. 2004. The engulfment process of programmed cell death in *Caenorhabditis elegans*. *Annu Rev Cell Dev Biol* **20:** 193–221.
   A detailed review of the engulfment process in nematodes.

Park SY, Kang KB, Thapa N, Kim SY, Lee SJ, Kim IS. 2008. Requirement of adaptor protein GULP during stabilin-2-mediated cell corpse engulfment. *J Biol Chem* **283:** 10593–10600.

Kinchen JM, Ravichandran KS. 2010. Identification of two evolutionarily conserved genes regulating processing of engulfed apoptotic cells. *Nature* **464:** 778–782.
   Two additional genes involved in engulfment of dying cells. Although not discussed in this chapter, the findings are of interest.

Waste Management

Kiss RS, Elliott MR, Ma Z, Marcel YL, KS Ravichandran. 2006. Apoptotic cells induce a phosphatidylserine-dependent homeostatic response from phagocytes. *Curr Biol* **16:** 2252–2258.

A-Gonzalez N, Bensinger SJ, Hong C, Beceiro S, Bradley MN, Zelcer N, Deniz J, Ramirez C, Diaz M, Gallardo G, et al. 2009. Apoptotic cells promote their own clearance and immune tolerance through activation of the nuclear receptor LXR. *Immunity* **31:** 245–258.
   In addition to illustrating the role of LXR in clearance of dying cells, this paper shows that activation of LXR may suppress disease consequences that arise due to defective apoptosis.

Dying Cells and Innate Immunity

Rock KL, Latz E, Ontiveros F, Kono H. 2010. The sterile inflammatory response. *Annu Rev Immunol* **28:** 321–342.

Birge RB, Ucker DS. 2008. Innate apoptotic immunity: The calming touch of death. *Cell Death Differ* **15:** 1096–1102.

Dying Cells and Adaptive Immunity

Green DR, Ferguson T, Zitvogel L, Kroemer G. 2009. Immunogenic and tolerogenic cell death. *Nat Rev Immunol* **9:** 353–363.
   A survey of the effects of dying cells on the adaptive immune system.

Gaipl US, Munoz LE, Grossmayer G, Lauber K, Franz S, Sarter K, Voll RE, Winkler T, Kuhn A, Kalden J, et al. 2007. Clearance deficiency and systemic lupus erythematosus (SLE). *J Autoimmun* **28:** 114–121.
   An overview of the possible role of engulfment defects in autoimmune disease.

Bianchi ME 2007. DAMPs, PAMPs and alarmins: All we need to know about danger. *J Leukoc Biol* **81:** 1–5.

Peng Y, Martin DA, Kenkel J, Zhang K, Ogden CA, Elkon KB. 2007. Innate and adaptive immune response to apoptotic cells. *J Autoimmun* **29:** 303–309.

Matzinger P. 1994. Tolerance, danger, and the extended family. *Annu Rev Immunol* **12:** 991–1045.
   The original review that triggered the idea that dying cells influence the adaptive immune response.

Skoberne M, Beignon AS, Larsson M, Bhardwaj N. 2005. Apoptotic cells at the crossroads of tolerance and immunity. *Curr Top Microbiol Immunol* **289:** 259–292.

Compensatory Proliferation

Fan Y, Bergmann A. 2008. Apoptosis-induced compensatory proliferation. The cell is dead. Long live the cell! *Trends Cell Biol* **18:** 467–473.

Ryoo HD, Gorenc T, Steller H. 2004. Apoptotic cells can induce compensatory cell proliferation through the JNK and the Wingless signaling pathways. *Dev Cell* **7:** 491–501.
   Two pathways involved in compensatory proliferation in flies.

Li F, Huang Q, Chen J, Peng Y, Roop DR, Bedford JS, CY Li. 2010. Apoptotic cells activate the "phoenix rising" pathway to promote wound healing and tissue regeneration. *Sci Signal* **3:** ra13.
   A possible mechanism of compensatory proliferation in response to apoptosis in mammals.

CHAPTER 10

Conradt B. 2009. Genetic control of programmed cell death during animal development. *Annu Rev Genet* **43:** 493–523.

Baehrecke EH. 2002. How death shapes life during development. *Nat Rev Mol Cell Biol* **3:** 779–787.

Meier P, Finch A, Evan G. 2000. Apoptosis in development. *Nature* **407:** 796–801.

Cell Death in Nematode Development

Ellis RE, Yuan JY, Horvitz HR. 1991. Mechanisms and functions of cell death. *Annu Rev Cell Biol* **7:** 663–698.
   An early survey of cell death in nematode development.

Ellis HM, Horvitz HR. 1986. Genetic control of programmed cell death in the nematode C. elegans. *Cell* **44:** 817–829.
   The original paper defining the genetic basis for cell death in nematode development.

Maurer CW, Chiorazzi M, Shaham S. 2007. Timing of the onset of a developmental cell death is controlled by transcriptional induction of the C. elegans ced-3 caspase-encoding gene. *Development* **134:** 1357–1368.

Abraham MC, Lu Y, Shaham S. 2007. A morphologically conserved nonapoptotic program promotes linker cell death in Caenorhabditis elegans. *Dev Cell* **12:** 73–86.
    Cell death in the tail-spike cell.

Ellis RE, Horvitz HR. 1991. Two C. elegans genes control the programmed deaths of specific cells in the pharynx. *Development* **112:** 591-603.
    The identification of CES1 and CES2 in specifying cell death in nematode development.

Metzstein MM, Hengartner MO, Tsung N, Ellis RE, Horvitz HR. 1996. Transcriptional regulator of programmed cell death encoded by Caenorhabditis elegans gene ces-2. *Nature* **382:** 545–547.

Zarkower D, Hodgkin J. 1992. Molecular analysis of the C. elegans sex-determining gene tra-1: A gene encoding two zinc finger proteins. *Cell* **70:** 237–249.

Cell Death in Fly Development

Hay BA, Guo M. 2006. Caspase-dependent cell death in Drosophila. *Annu Rev Cell Dev Biol* **22:** 623–650.

Baehrecke EH. 2000. Steroid regulation of programmed cell death during Drosophila development. *Cell Death Differ* **7:** 1057–1062.

Rusconi JC, Hays R, Cagan RL. 2000. Programmed cell death and patterning in Drosophila. *Cell Death Differ* **7:** 1063–1070.

Berry DL, Baehrecke EH. 2007. Growth arrest and autophagy are required for salivary gland cell degradation in Drosophila. *Cell* **131:** 1137–1148.
    The first clear-cut description of autophagy-dependent cell death in development.

Denton D, Shravage B, Simin R, Mills K, Berry DL, Baehrecke EH, Kumar S. 2009. Autophagy, not apoptosis, is essential for midgut cell death in Drosophila. *Curr Biol* **19:** 1741–1746.
    Another example of developmental cell death dependent on autophagy.

Cell Death in Vertebrate Development

Montero JA, Hurlé JM. 2010. Sculpturing digit shape by cell death. *Apoptosis* **15:** 365–375.

Coucouvanis E, Martin GR. 1995. Signals for death and survival: A two-step mechanism for cavitation in the vertebrate embryo. *Cell* **83:** 279–287.

Ishizuya-Oka A, Hasebe T, Shi YB. 2010. Apoptosis in amphibian organs during metamorphosis. *Apoptosis* **15:** 350–364.

Cell Death and Selection

Yuen EC, Howe CL, Li Y, Holtzman DM, Mobley WC. 1996. Nerve growth factor and the neurotrophic factor hypothesis. *Brain Dev* **18:** 362–368.

O'Leary DD, Fawcett JW, Cowan WM. 1986. Topographic targeting errors in the retinocollicular projection and their elimination by selective ganglion cell death. *J Neurosci* **6:** 3692–3705.
    An example of how cell death can select for proper neuronal connections.

Palmer E. 2003. Negative selection—Clearing out the bad apples from the T-cell repertoire. *Nat Rev Immunol* **3:** 383–391.

Strasser A, Puthalakath H, O'Reilly LA, Bouillet P. 2008. What do we know about the mechanisms of elimination of autoreactive T and B cells and what challenges remain. *Immunol Cell Biol* **86:** 57–66.

Bouillet P, Purton JF, Godfrey DI, Zhang LC, Coultas L, Puthalakath H, Pellegrini M, Cory S, Adams JM, Strasser A. 2002. BH3-only Bcl-2 family member Bim is required for apoptosis of autoreactive thymocytes. *Nature* **415:** 922–926.
  The role of BIM in the negative selection of immature T cells.

CHAPTER 11

Cancer and Apoptosis

Hanahan D, Weinberg RA. 2000. The hallmarks of cancer. *Cell* **100:** 57–70.
  An influential overview of how cancer occurs, including the role of apoptosis evasion.

Green DR, Evan GI. 2002. A matter of life and death. *Cancer Cell* **1:** 19–30.
  A proposal that cancer is a largely a consequence of unregulated proliferation combined with evasion of apoptosis.

Letai AG. 2008. Diagnosing and exploiting cancer's addiction to blocks in apoptosis. *Nat Rev Cancer* **8:** 121–132.
  Harnessing our knowledge of cell death in treating cancer.

Pelengaris S, Khan M, Evan GI. 2002. Suppression of Myc-induced apoptosis in β cells exposes multiple oncogenic properties of Myc and triggers carcinogenic progression. *Cell* **109:** 321–334.
  An important paper that illustrates the requirements of c-Myc–induced cancer.

p53

Junttila MR, Evan GI. 2009. p53—A Jack of all trades but master of none. *Nat Rev Cancer* **9:** 821–829.

Vousden KH, Prives C. 2009. Blinded by the light: The growing complexity of p53. *Cell* **137:** 413–431.

Vousden KH, Lane DP. 2007. p53 in health and disease. *Nat Rev Mol Cell Biol* **8:** 275–283.

Caspari T. 2000. How to activate p53. *Curr Biol* **10:** R315–R317.

Sherr CJ. 2006. Divorcing ARF and p53: An unsettled case. *Nat Rev Cancer* **6:** 663–673.
  A discussion of the functions of ARF in p53 activation and other possible roles.

Murray-Zmijewski F, Slee EA, Lu X. 2008. A complex barcode underlies the heterogeneous response of p53 to stress. *Nat Rev Mol Cell Biol* **9:** 702–712.

Oren M, Levine AJ. 1983. Molecular cloning of a cDNA specific for the murine p53 cellular tumor antigen. *Proc Natl Acad Sci* **80:** 56–59.
  One of two early papers describing a role for p53 in cancer.

Lane DP. 1984. Cell immortalization and transformation by the *p53* gene. *Nature* **312:** 596–597.
  See above. Interestingly, the p53 cloned was a dominant-negative mutant that promoted (rather than suppressed) cellular transformation.

p53 and Apoptosis

Yu J, Zhang L. 2008. PUMA, a potent killer with or without p53. *Oncogene* (suppl 1) **27:** S71–S83.
  A review covering the roles for PUMA as a mediator of p53-induced apoptosis.

Green DR, Kroemer G. 2009. Cytoplasmic functions of the tumour suppressor p53. *Nature* **458:** 1127–1130.

Tumor Suppression

Lowe SW, Cepero E, Evan G. 2004. Intrinsic tumour suppression. *Nature* **432:** 307–315.

Christophorou MA, Ringshausen I, Finch AJ, Swigart LB, Evan GI. 2006. The pathological response to DNA damage does not contribute to p53-mediated tumour suppression. *Nature* **443:** 214–217.
 An exploration of how p53 suppresses oncogenesis.

Campisi J. 2005. Senescent cells, tumor suppression, and organismal aging: Good citizens, bad neighbors. *Cell* **120:** 513–522.
 An overview of tumor suppression, beyond apoptosis.

Autophagy and Cancer

White E, Karp C, Strohecker AM, Guo Y, Mathew R. 2010. Role of autophagy in suppression of inflammation and cancer. *Curr Opin Cell Biol* **22:** 212–217.

Mathew R, Karantza-Wadsworth V, White E. 2007. Role of autophagy in cancer. *Nat Rev Cancer* **7:** 961–967.

Degenhardt K, Mathew R, Beaudoin B, Bray K, Anderson D, Chen G, Mukherjee C, Shi Y, Gelinas C, Fan Y, et al. 2006. Autophagy promotes tumor cell survival and restricts necrosis, inflammation, and tumorigenesis. *Cancer Cell* **10:** 51–64.
 Evidence that autophagy acts as a tumor suppressor.

Maclean KH, Dorsey FC, Cleveland JL, Kastan MB. 2008. Targeting lysosomal degradation induces p53-dependent cell death and prevents cancer in mouse models of lymphomagenesis. *J Clin Invest* **118:** 79–88.
 Induction of autophagic cell death by chloroquine, applied to cancer therapy.

CHAPTER 12

Modeling Apoptosis

Aldridge BB, Burke JM, Lauffenburger DA, Sorger PK. 2006. Physicochemical modelling of cell signalling pathways. *Nat Cell Biol* **8:** 1195–1203.

Spencer SL, Gaudet S, Albeck JG, Burke JM, Sorger PK. 2009. Non-genetic origins of cell-to-cell variability in TRAIL-induced apoptosis. *Nature* **459:** 428–432.
 Evidence of stochastic processes in apoptosis.

Pomerening JR. 2008. Uncovering mechanisms of bistability in biological systems. *Curr Opin Biotechnol* **19:** 381–388.
 An introduction to bistability.

Cui J, Chen C., Lu H, Sun T, Shen P. 2008. Two independent positive feedbacks and bistability in the Bcl-2 apoptotic switch. *PLoS One* **3:** e1469.
 This and the following paper model BCL-2 family interactions in apoptosis.

Sun T, Lin X, Wei Y, Xu Y, Shen P. 2010. Evaluating bistability of Bax activation switch. *FEBS Lett* **584:** 954–960.
 See above.

Legewie S, Bluthgen N, Herzel H. 2006. Mathematical modeling identifies inhibitors of apoptosis as mediators of positive feedback and bistability. *PLoS Comput Biol* **2:** e120.

Making Cells Die

Straathof KC, Pule MA, Yotnda P, Dotti G, Vanin EF, Brenner MK, Heslop HE, Spencer DM, Rooney CM. 2005. An inducible caspase 9 safety switch for T-cell therapy. *Blood* **105:** 4247–4254.
   One way of triggering apoptosis, on demand.

Chonghaile TN, Letai A. 2008. Mimicking the BH3 domain to kill cancer cells. *Oncogene* (suppl 1) **27:** S149–S157.
   A survey of BCL-2 inhibitors, including ABT-737, and how they work.

Oltersdorf T, Elmore SW, Shoemaker AR, Armstrong RC, Augeri DJ, Belli BA, Bruncko M, Deckwerth TL, Dinges J, Hajduk PJ, et al. 2005. An inhibitor of Bcl-2 family proteins induces regression of solid tumours. *Nature* **435:** 677–681.
   The first description of ABT-737 and how it was discovered.

Brown CJ, Lain S, Verma CS, Fersht AR, Lane DP. 2009. Awakening guardian angels: Drugging the p53 pathway. *Nat Rev Cancer* **9:** 862–873.

Making Cells Live

Enari M, Hug H, Nagata S. 1995. Involvement of an ICE-like protease in Fas-mediated apoptosis. *Nature* **375:** 78–81.
   The first description of the protective effect of caspase (called "ICE-like protease) inhibitors in protection from liver destruction induced by injection of antibodies to CD95 (called "Fas").

Degterev A, Hitomi J, Germscheid M, Ch'en IL, Korkina O, Teng X, Abbott D, Cuny GD, Yuan C, Wagner G, et al. 2008. Identification of RIP1 kinase as a specific cellular target of necrostatins. *Nat Chem Biol* **4:** 313–321.

He S, Wang L, Miao L, Wang T, Du F, Zhao L, Wang X. 2009. Receptor interacting protein kinase-3 determines cellular necrotic response to TNF-α. *Cell* **137:** 1100–1111.
   The role of RIPK3 (and RIP-dependent necrosis) in different cell death phenomena.

The Button Experiment

Kauffman S. 1995. *At home in the universe.* Oxford University Press, Oxford.
   The first description of the button experiment, a simple idea with intriguing consequences.

Spierings D, McStay G, Saleh M, Bender C, Chipuk J, Maurer U, Green DR. 2005. Connected to death: The (unexpurgated) mitochondrial pathway of apoptosis. *Science* **310:** 66–67.
   An introduction to an extensive online survey of many different proteins that have been described to impact on the mitochondrial pathway of apoptosis.

Index

A1, 66
ABC1, 136
ABC7, 136
ABT-737, 184–185
Acinus, caspase cleavage, 22
Adaptive immunity, impact of dying cells, 144–146
Adenosine nucleotide transporter (ANT), 56
AIF. See Apoptosis-inducing factor
AIM2 inflammasome, 105
AKT, 22, 77, 122, 164, 174
AMPK, 122
Annexin V, 130
Anoikis, 76–77
ANT. See Adenosine nucleotide transporter
Antagonistic pleiotropy, 164
Anthrax toxin, 105
AP-1, 93
APAF1, 49–51, 54, 57–58, 60, 78–79, 106–107, 155
APO-1. See CD95
Apoptosis
   bottom-up view, 8–9
   clearance of cells. See Clearance, dead cells
   definition, 6
   inducers, 12–13
   pathways. See Death receptor pathway; Mitochondrial pathway
Apoptosis-inducing factor (AIF), 55
ARK, 58–59, 152
ASC, 100–101
ATG5, 120–121, 124, 153
ATG7, 153
ATG10, 121
ATG12, 120–121, 124
ATG13, 120–121
ATG16, 121

Autophagic cell death
   definition, 6, 123
   overview, 118–119
Autophagy
   cancer, 175–176
   mitochondria. See Mitophagy
   pathway, 120–123
   survival mechanism, 119–120

Baculovirus IAP repeat (BIR), 42–43
BAD, 74–75, 83
BAI1, 134
BAK, 64–73, 83–84, 125, 154–155, 158–171
BAX, 64–66, 68–73, 83–84, 125, 154–155, 158, 170
BCL-2 proteins. See also specific proteins
   anti-apoptotic proteins, 66–68, 77–78
   autophagy regulation, 122
   bacterial toxins in origin, 81–82
   BH domains, 63–64
   BH3-only proteins, 68–73
   cancer modulation, 169–171
   classification, 9
   invertebrates, 78–79
   miscellaneous functions, 82–86
   mitochondrial function, 83–86
   phosphorylation, 77–78
   structural homology, 81
   therapeutic targeting, 184–185
   types, 63–64
   viral proteins, 80–81
BCL-xL, 66, 68, 75, 77–78, 81, 173
Beclin-1, 85–96, 120–121, 175
BID, 70, 73–74, 95–97, 170
BIM, 70–71, 74–76, 154–155, 160–161
BIR. See Baculovirus IAP repeat
Bistability in cell death pathways, 180–182

BMF, 75–76, 154–155
BMPs. *See* Bone morphogenic proteins
BNIP, 84–85
BOK, 65, 173
Bone morphogenic proteins (BMPs), 154
BR-C, 152–153
Buffy, 79, 84

C1q, 132–133, 135
C3b, 133, 135
CAD. *See* Caspase-activated DNase
E-Cadherin, caspase cleavage, 22
Calreticulin, 129, 135–136, 145
Cancer
    apoptosis and proliferation, 163–166
    apoptosis as Achilles heel, 174–175
    autophagy, 175–176
    frequency, 163
    p53
        activation, 167–168
        cytosolic protein function in apoptosis, 171–172
        DNA-damage induction, 168
        invertebrate apoptosis, 172–173
        mutation and cancer susceptibility, 168–169
        target genes, 169–171
    tumor-suppressor proteins, 166, 173–174
CARD. *See* Caspase-recruitment domain
Caspase. *See also specific caspases*
    activation
        death folds and adapter protein interactions, 34–35
        executioner caspases, 29–31
        initiator caspases, 31–34, 36–37
        mitochondrial activation, 49–51
    domains, 17–18
    inhibitors
        endogenous inhibitors, 40
        synthetic inhibitors, 41
        viral inhibitors, 42–45
    invertebrates, 18
    mechanism of action, 15–16
    membrane blebbing, 25–26
    mitochondrial effects, 26
    substrate specificity, 19–23
    types, 16–17
Caspase-1
    activation by inflammasomes
        AIM2 inflammasome, 105
        ASC role, 100–101
        IPAF inflammasome, 103–104

        NLRP1 inflammasome, 105
        NLRP3 inflammasome, 103–104
        Nod-like receptors, 101–102
        overview, 99–100
        Toll-like receptor role, 101
    functional overview, 38–39
    mechanism of cell death, 106
Caspase-2
    activation, 108–109
    function, 39
Caspase-3
    activation, 30–31
    inhibition, 41
    knockout mouse, 28
    substrate specificity, 21
Caspase-6, activation, 30
Caspase-7
    activation, 30–31, 106
    knockout mouse, 28
Caspase-8
    activation by dimerization, 37
    inhibition, 45
    necroptosis inhibition, 118
Caspase-9
    activation by dimerization, 36
    knockout mouse, 28, 51, 155
Caspase-12, 38, 40
Caspase-activated DNase (CAD), 24–26, 141
Caspase-independent cell death (CICD)
    mechanisms, 55
    mitochondrial outer membrane permeabilization role, 54
Caspase-recruitment domain (CARD), 35, 37, 49, 58, 99–100, 108
β-Catenin, caspase cleavage, 22
CD31, 131
CD36, 135–136
CD47, 130–131
CD68, 135
CD95
    apoptosis induction mechanism, 89–92
    hepatocyte mitochondrial pathway of apoptosis, 97–98
    induction by p53, 170
    knockout mouse, 92, 116
    ligands, 87
    signaling defects and lymphoproliferative disease, 92
CDK1, 125, 136–138
CED2, 139

CED3, 18, 27, 37, 57, 150–151
CED4, 58–59, 78, 84, 150–151
CED5, 139
CED6, 138
CED7, 136–137
CED9, 58, 78–79, 84, 151
CED10, 138
CED12, 139
CEP1, 173
CES1, 150–151
CES2, 150
CICD. See Caspase-independent cell death
Clearance, dead cells
    adaptive immunity impact of dying cells, 144–146
    autoimmune disease implications, 146–147
    bind-me signals on dying cells
        bridging molecules, 130–132
        invertebrates, 135–138
        overview, 129–130
        phagocyte receptors, 133–135
    clear-me process, 141–142
    compensatory proliferation, 147–148
    don't-eat-me signals, 130–131
    eat-me signals, 138–141
    find-me signals, 128–129
    innate immunity impact of dying cells, 142–143
    overview, 127–128
    tether and tickle model, 133
Combinatorial peptide library, caspase substrate specificity studies, 20–21
Compensatory proliferation, 147–148
Cornification, definition, 8
CRKII, 140
CrmA, 45
Croquemort, 136–137, 153
Cyclophilin D, 115, 187
Cyclosporin A, 115–116, 187
Cytochrome c, 50–52
Cytotoxic T lymphocyte, apoptosis induction, 31–32

Damage-associated molecular patterns (DAMPs), 100–101, 128–129, 143–145
Damm, 18
DAMPs. See Damage-associated molecular patterns
DCP1, 18
DD. See Death domain
Death domain (DD), 35, 88–90, 93–94
Death effector domain (DED), 35

Death receptor pathway. See also specific receptors
    CD95, 89–92
    invertebrates, 98
    mitochondrial pathway involvement, 95–98
    necrosis role, 116–118
    overview, 11
    receptors and ligands, 87–89
    TNFR1 signaling, 93–95
    TRAIL-induced apoptosis, 93
Death-inducing signaling complex (DISC), 90–91
DeBCL, 79, 84
Decapentaplegic (Dpp), 147
DED. See Death effector domain
Development, apoptosis role
    cell selection
        neurons, 157–158
        T cells, 158–161
    Drosophila metamorphosis, 152–154
    nematode development, 149–152
    vertebrate development, 154–157
dFADD, 98
DFF45. See Caspase-activated DNase
Diablo. See Smac
DIAP1, 42, 44, 58–60, 152
DISC. See Death-inducing signaling complex
DNA damage
    autophagy induction, 123
    cell death induction, 113
    p53 induction, 168
DOCK180, 139–140
Dpp. See Decapentaplegic
DR4, TRAIL, 87
DR5
    induction by p53, 170
    TRAIL, 87
DR6, 87
DRAM, 124
Draper, 138, 153
Dredd, 37, 98
Drice, 18, 152
Dronc, 27, 37, 42, 57–58, 60, 152
Drosophila metamorphosis, cell death, 152–154
DRP-1, 84

E74, 152–153
E93, 152–153
Ecdysone, 153
EGL1, 79, 84, 150–151
ELMO, 139–140
Endonuclease G, 55

ERK, 75
Excitotoxicity, necrosis induction, 113–114
Executioner caspases
    activation, 29
    mitochondrial effects, 26
    substrates, 23
    types, 16

FADD, 89–92, 94, 124
Fas. See CD95
FIP200, 121
FK506, 183–184
FKB12, 183
FLIP
    caspase-8 interactions in activation, 37
    CD95 apoptosis induction mechanism, 91–92
    necroptosis inhibition, 118
FOXO3a, 74–75

Gas6, 131, 134
GATA-1, caspase cleavage, 22
Gelsolin, caspase cleavage, 22
Gene therapy, 183
Granzyme B, apoptosis induction, 31–32
GSK3, 77, 187–188
GULP, 138

Hermaphrodite-specific neuron (HSN), 150
Hid, 152
HIF-1, 85
HLH2, 150
HLH3, 150
HMGB1, 145
Horvitz, Robert, ix, 1
HRK, 157
HSN. See Hermaphrodite-specific neuron
HtrA2. See Omi

IAP. See Inhibitor of apoptosis protein
iCAD, 22, 24–25, 141
IL-33, caspase cleavage, 22
Immunogenic apoptosis, 145
Inflammasome. See Caspase-1
Inflammatory caspases
    function in secretion and cell death, 38–39
    overview, 17
Inhibitor of apoptosis protein (IAP), 42–44, 53, 60, 182
Innate immunity, impact of dying cells, 142–143
Initiator caspases
    activation, 32–34, 36–37
    overview, 16

IP3 receptor, BCL-2 modulation, 83
IPAF inflammasome, 103–104
Ischemia/reperfusion injury
    mitochondria role, 115–116
    potassium and calcium changes, 114s

JNK, 75, 93–94, 160

Lamin A/C, caspase cleavage, 22
Lamin B1, caspase cleavage, 22
LC3, 120–121
LDL. See Low-density lipoprotein
Li-Fraumeni syndrome, 168–169
Low-density lipoprotein (LDL), oxidation, 135
LOX1, 135
LPA. See Lysophosphatidic acid
LRP1, 135–136, 138
Lysophosphatidic acid (LPA), 128

MALT, caspase-8 interactions in activation, 37
Mannose-binding lectin (MBL), 132
MAPK. See Mitogen-activated protein kinase
MBL. See Mannose-binding lectin
MCL-1, 66, 68, 69, 77, 173
MDM2, 181, 185
MER, 134
MFG-E8. See Milk fat globulin-E8
Milk fat globulin-E8 (MFG-E8), 131, 134
Mitochondrial outer membrane permeabilization (MOMP), 10–11, 51–52, 54, 57, 63–66, 95–97, 174–175, 180–181
Mitochondrial pathway
    caspase activation, 49–51
    caspase-independent cell death, 54–55
    death receptor pathway involvement, 95–98
    Just So Story of mitochondria and cell death, 47–48, 81–82, 106–107
    overview, 9–11
Mitochondrial permeability transition (MPT), 55–57, 115, 187
Mitogen-activated protein kinase (MAPK), 75
Mitophagy, BCL-2 role, 84–86
Mitotic catastrophe, 111, 124–125
MOMP. See Mitochondrial outer membrane permeabilization
MPT. See Mitochondrial permeability transition
Mst1, caspase cleavage, 22
mTOR, 122
MULE, 77
Myc, 163–165, 171

NADH, depletion, 113
NADPH oxidase, 113
NAIP5, 105
NDUFS1, 22, 26, 145
Necroptosis, 116–118
Necrosis
   apoptosis comparison, 112
   death receptor role, 116–118
   definition, 7
   excitotoxicity, 113–114
   ischemia/reperfusion injury, 114–116
   PARP role, 112–113
   secondary necrosis, 112
Necrostatins, 117
Nematode development, cell death, 149–152
NEMO, 93
Neuron, apoptosis in selection, 157–158
NF-κB. See Nuclear factor-κB
NIX, 84–85
NLRP1 inflammasome, 105
NLRP3 inflammasome, 103–104
NLRs. See Nod-like receptors
Nod-like receptors (NLRs), 101–107, 143
NOXA, 170
Nuclear factor-κB (NF-κB), 93–94

Omi, 53, 60–61

P2X$_7$, 100–101
p35, baculovirus, 45
p53
   activation, 113, 167–168
   cytosolic protein function in apoptosis, 171–172
   DNA-damage induction, 168
   invertebrate apoptosis, 172–173
   mutation and cancer susceptibility, 168–169
   target genes, 169–171
PAK-2, caspase cleavage, 22, 25
PAMPs. See Pathogen-associated molecular patterns
PARP. See Poly-ADP-ribose polymerase
Pathogen-associated molecular patterns (PAMPs), 100–101, 143
Pattern-recognition receptors (PRRs), 143
Permeability transition pore (PTP), 55–57, 115
Peroxisome proliferator-activated receptor γ (PPARγ), 142–143, 147
Phagocytosis. See Clearance, dead cells
Phagosome, 141
Phosphatidylserine
   annexin V for detection, 130

   bind-me signal on dying cells, 129
   bridging molecules, 131–133
   receptors, 134
Phosphoinositide-3-kinase (PI3K), 120–121
Phospholipase A$_2$, 147
PI3K. See Phosphoinositide-3-kinase
PIDD, 108–109
PIDDosome, 108–109
PIP$_3$, 174
Poly-ADP-ribose polymerase (PARP)
   caspase cleavage of PARP-1, 22
   necrosis role, 112–113
Protein-S, 131
PRRs. See Pattern-recognition receptors
PTEN, 174
PTP. See Permeability transition pore
PUMA, 70, 170, 172
PyD. See Pyrin domain
Pyrin domain (PyD), 35, 100
Pyroptosis, 111

RAC1, 138–140
RAIDD, 108
Reaper, 152
Relish, 98
RHOG, 140
RIP-1, caspase cleavage, 22
RIPK1, 93–94, 117–118, 186
RIPK3, 117, 186
ROCK-1, caspase cleavage, 22, 25

Scavenger receptors, 135
SIRP1α, 131
Smac, 52–53, 60–61
SOCS1, 143
Stabilin-2, 134
Surfactant proteins, 132
Systemic lupus erythematosis, 146–147

T cell
   apoptosis in selection, 158–161
   cytotoxic T lymphocyte apoptosis induction, 31–32
TCTP. See Translationally controlled tumor protein
TGF-β. See Transforming growth factor-β
Thrombospondin-1, 131
TIM4, 134
TLRs. See Toll-like receptors
TNF-receptor associated factor-2 (TRAF-2), 93
TNFR1
   apoptotic signaling, 93–95, 98
   ligands, 87

Tolerogenic apoptosis, 145
Toll-like receptors (TLRs), 101, 143
TORC1, 122
TORC2, 122
TRADD, 93–94, 117
TRAF, 93–94, 117
TRAF-1, caspase cleavage, 22
TRAF-2. *See* TNF-receptor associated factor-2
TRAIL receptor-1. *See* DR4
TRAIL receptor-2. *See* DR5
Transforming growth factor-β (TGF-β), 143, 146
Translationally controlled tumor protein (TCTP), 168
TRIO, 140

Tumor necrosis factor receptors. *See* Death receptor pathway

ULK1, 121

VDAC. *See* Voltage-dependent anion channel
Vertebrate development, cell death, 154–157
Vimentin, caspase cleavage, 22
Voltage-dependent anion channel (VDAC), 56
VPS34, 120–121

WD domain, 50, 58, 60

XIAP, 22, 43–44, 53, 61, 96–98